100 TOP
GAMES APPS
MADE EASY

ANDROID • iPHONE • iPAD
GOOGLE • AMAZON FIRE • SAMSUNG

This is a FLAME TREE book
First published 2013

Publisher and Creative Director: Nick Wells
Project Editor: Polly Prior
Art Director: Mike Spender
Layout Design: Jane Ashley
Digital Design and Production: Chris Herbert
Copy Editor: Anna Groves
Technical Editor: Mark Mayne
Screenshots: Julian Richards and Chris Smith
Special thanks to: Laura Bulbeck, Esme Chapman and Emma Chafer

This edition first published 2013 by
FLAME TREE PUBLISHING
Crabtree Hall, Crabtree Lane
Fulham, London SW6 6TY
United Kingdom

www.flametreepublishing.com

© 2013 this edition Flame Tree Publishing

ISBN 978-0-85775-814-9

Printed in China

All non-screenshot pictures are © the following: 9 Gina Weakley Johnson; 10l 2013 Apple Inc.; 10r 1995–2013 SAMSUNG; and courtesy of Shutterstock and © the following contributors: 18 chert28; 28 Dudarev Mikhail; 84 Thomas Cristofoletti; 140 somchaij ; 196 MeiKIS.

100 TOP
GAMES APPS
MADE EASY

ANDROID • iPHONE • iPAD
GOOGLE • AMAZON FIRE • SAMSUNG

JULIAN RICHARDS AND CHRIS SMITH

**FLAME TREE
PUBLISHING**

CONTENTS

Get ready for some adrenaline-charged, action-packed, hand-held gaming! In these titles, you'll be leaping through the jungle, escaping from temples, reclaiming stolen eggs, single-handedly fighting off battalions of alien invaders, putting medieval foes to the sword with sweet combination blows, slashing through fruit like a ninja and collecting rings to your heart's content. (Chris Smith)

PUZZLE & TRIVIA GAMES

These games will test your powers of logical thought, word skills and quiz knowledge to the limit. There are reimagined versions of classic board games, addictive physics-based problem-solving games, enough trivia games to earn you a spot on Mastermind and more than a handful of card games. Here, you'll learn how to play Sudoku without having to scribble over your answers, while playing Scrabble with friends hundreds of miles away. (Julian Richards)

SPORTS & RACING GAMES

Feel like managing your favourite team to Premier League glory, or even scoring the winning goal yourself? Then there are plenty of sports simulations to help you achieve your virtual dreams. Try your hand at becoming a tennis, golf or even darts pro – the only equipment required is a mobile handset. We've also amassed the finest racing and automotive games to feed your need for speed. (Chris Smith)

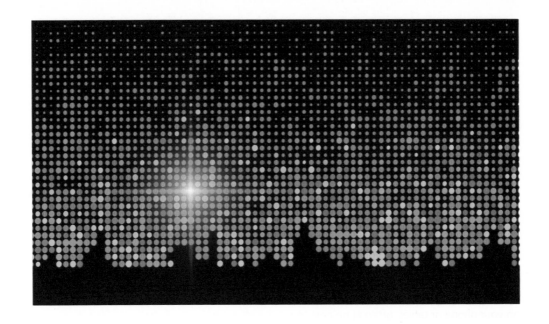

SIMULATION & STRATEGY GAMES

If you feel like investing a little time in your gaming activities, this genre will allow you to build your own city, restaurant, army and more. You can even rebuild The Simpsons' hometown of Springfield in your own images. There are wars to be waged, planets to be colonized, towers to defend and wonderful characters to invent. Start building your empire today. (Julian Richards)

INTRODUCTION

Welcome to *100 Top Games Apps Made Easy* for Android and iOS phones and tablets. Whether you're a virtual-racing speedster, would-be Empire builder, card shark, quiz brain, or you just want to fling some Angry Birds at smirking, egg-stealing pigs, then this book has something for you.

By introducing you to the finest mobile games on offer, we hope to bring an end to the tedium of commutes to work, or help while away the tiresome hours waiting for appointments and uneventful evenings sitting in front of the telly. With the help of these 100 mobile gaming gems, you'll never be bored again.

We'll bring you the 100 best games, from a variety of genres, which you can easily get hold of on your phone or tablet, often at no cost to you whatsoever. You're entering a whole new world of exciting games, all at your fingertips and instantly obtainable. Are you excited? We're excited for you!

Above: Games on mobile devices have come a long way.

MOBILE GAMES

Most of us have been playing games on our mobile phones for years, although looking back on them now they will seem positively primitive. Anyone over the age of 25 will remember Snake on some of the old Nokia handsets. Although chasing dots of 'food' while trying to avoid crashing into your own 'tail' was fun and addictive, thankfully things have become a lot more interesting since then.

BETTER GADGETS MEANS BETTER GAMES

Most smartphones and tablets made by Apple, Samsung, ASUS and others have beautiful high-resolution screens, along with high-powered computing and graphics chips, meaning that your iOS and Android devices have become powerful hand-held gaming machines, capable of competing with dedicated hand-held consoles like the Sony PS Vita and Nintendo 3DS. Better still, the games cost a fraction of the price!

THE TOUCHSCREEN REVOLUTION

The first colour screens brought mobile games out of monochrome monotony, but it was the advent of touchscreens that allowed mobile gaming to become the phenomenon it is today.

The vast majority of devices have incredibly sensitive and accurate multi-touch screens, which means you can use multiple fingers at the same time to push, flick, grab, drag, swipe or pan. In some games, the directional and action buttons you see on traditional games controllers are replaced by on-screen keys.

SMARTPHONES AND TABLETS

An ever-increasing number of people now own smartphones such as the Apple iPhone, which runs the iOS software, or handsets on Google's Android platform. Additionally, more and more people are buying tablets like the iPad and the Google Nexus 7, which offer larger screens and more precise controls for even better gaming.

Above: The Google Nexus 7

Apple iOS Gadgets

If you own an iPhone, iPod Touch or an iPad, then that quite lovely piece of kit has Apple's iOS software on board. Whether your device is old or new, you'll be able to access over 100,000 games through a magical portal to another universe, otherwise known as the App Store.

Google Android Gadgets

If you have a phone or tablet made by Samsung, HTC, Acer, Motorola or LG (to name but a few), it's probably running the Android operating system made by Google. To get your games, hit the Play Store icon.

Above: Access the App Store through the Apple iPhone 5.

HOW TO DOWNLOAD GAMES

iOS and Android devices don't come with any games, and you can't load games by inserting cartridges or discs, so how do you get the darned things on there? Through the internet, of course! As you'd expect, all titles are available instantly, usually taking less than a minute to download over Wi-Fi.

Above: Phones by makers such as Samsung will probably run the Android operating system.

Downloading iOS Games

Firstly, hit the App Store icon on your device; it's blue and has an 'A' made out of a pencil, a paintbrush and a ruler. Hit Categories in the top-left corner and select Games from the menu.

You can then select All Games to go to the Games homepage or select a category (such as Action, Sports). You can also look for a particular title using the Search icon.

Once you've found your game, you'll see a page containing screenshots, reviews and a description. To install it, hit the green button that says 'FREE' or shows the price (for example, £0.69/$0.99), and then select 'INSTALL' or 'BUY APP'. You'll need to enter your Apple ID password. You should have opened an account when you set up your device. If not, go to www.apple.com/support

You'll then be returned to your phone's home screen, where the app will be shown downloading and installing. Once the download is complete, hit the game's icon and you're ready to start playing.

Above: The Angry Birds app in the App Store; click on the price button to start the download process.

Downloading Android Games

Hit the Play Store icon on your Android smartphone or tablet and select the Games category on the home screen. From here you can browse genres and select games to install, or hit the Search icon at any time to look for a particular title.

If the app costs money, you'll see a price next to the app; if it's free, you'll just see the word 'Install'. When you begin the download, you'll see its progress in the notifications bar at the top of the screen. Once the installation is complete, you'll be notified and you can select the game to open and begin playing.

Download Size

The size of the game determines how long it takes to download and how much storage space it takes up on your phone. Each game reviewed in this book has been put into one of three categories:

- ◎ **Under 100MB = Small**
- ◎ **100MB–500MB = Medium**
- ◎ **500MB+ = Large**

Above: Modern Combat 4 has a large download size.

KEEPING IT CASUAL

As you've bought this book, we dare say you're probably not a big gamer, up until all hours playing World of Warcraft, or partaking in elaborate co-op missions with your online Call of Duty team. That's okay, neither are we! To us, mobile games are all about having a little fun, reliving classic titles and exploring new games, filling the odd gap in the day and enjoying playing with friends, either over the web or by passing handsets around.

FOUR CHAPTERS

This book will not be ranking the top 100 games in order, as that's hugely subjective and, with new releases hitting the virtual shelves every day, can change with the same frequency. Instead, this book will offer a guide to what we consider to be the finest and most varied games on offer in each of the following categories: Arcade & Adventure, Puzzle & Trivia, Sports & Racing, Simulation & Strategy.

Breaking Down Each Game

Each game featured within this book gets its own double-page spread, where we'll outline the premise and the main aims, and explain the story behind each one. You'll learn how to master the controls and pick up some tips along the way.

Staying In Control

Most games involve prodding, swiping, dragging and holding down or releasing areas on the touchscreen, but motion sensors in handsets also offer controls through tilting. For example, in Temple Run (see pages 74–75), you'll swipe to turn, swipe up to jump and tilt the handset left and right to steer your character out of danger. In Angry Birds (see pages 36–37), you press and hold the bird, drag back the slingshot, move up and down to perfect your aim, and release to fire!

Hot Tips

As we tested and read about these games, we picked up a few handy hints on how to improve performance. You'll see these Hot Tip boxes scattered throughout.

You Might Also Like

You will not fall in love with every game in this book. If you hate football, you're unlikely to embark on a season in charge of Barcelona on FIFA Soccer. However, if you are partial to a certain title, the likelihood is that there are plenty more like it for you to explore, so check out these related-titles boxes.

Above: Once the game has finished installing, you can begin playing.

CATEGORY ICONS

So that you can easily pick out which main subject category each game fits into, we have provided a handy icon which represents one of eight different themes. So, if you're in the mood for a strategy game for example, you'll have figured out that you should turn to the Strategy & Simulation Games chapter; but these icons will help you further identify those games that are mainly strategy within the chapter.

Below is a key to explain which category each icon represents:

Arcade

Adventure

Puzzle

Trivia

Sports

Racing

Strategy

Simulation

HOW WE CHOSE THE GAMES

We considered a number of factors in constructing the list, including:

- **Origin:** We expressed a preference for games designed for touchscreens rather than console games shoehorned into the app format.

- **History**: We also considered the popularity and history of a gaming series, such as Sonic the Hedgehog or FIFA.

- **Popularity**: We looked at user review scores as well as critical review scores.

- **Variety**: Rather than have multiple games of the same type, we picked the best game of that type.

- **Originality**: If a game brought unique gameplay, story, visuals, or even added a new twist to a tested formula, it was more likely to make the cut.

- **Availability**: We leaned towards games available for both Apple and Android devices, but there were some iOS-only games that were so good (such as Infinity

Above: Sonic the Hedgehog has a long history of being enjoyed by gamers and had to be included amongst the Top 100 Games Apps..

Above: Would Asphalt 7: Heat make your Top 100?

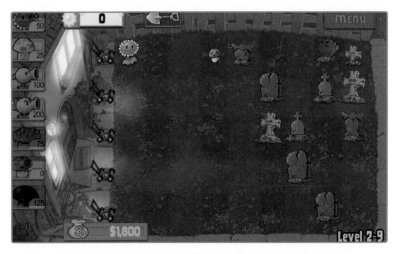

Blade) that they were impossible to ignore. In those circumstances, we have tried to suggest the best Android alternative.

○ **Personal taste**: Lastly, we chose these games because we love them and think you will too.

Above: Plants vs. Zombies is a fantastically fun twist on the tower-defence game and has taken the gaming world by storm.

COST

Each game fits into one of the following categories:

○ **Free**: You won't pay a penny, but may have to put up with adverts

○ **Free (with in-app purchases)**: Free to download and play, but spending real money unlocks new content

○ **Free/Paid**: There are both free and paid versions of the same game

○ **Paid**: You'll need to pay to download this game

Most of the 100 games in this book are free to download or cost less than a cup of coffee, but we've included some that we think you just can't afford to ignore. Some 'free' titles require in-app purchases to continue enjoying the game beyond a basic level. Don't worry, we'll alert you to those. In some free games, you can also pay a small fee to upgrade and get rid of annoying advertisements.

Above: You'll need to pay to download Tiny Wings.

ABOUT THE AUTHORS

Chris Smith writes about mobile gaming for a number of the UK's leading technology publications. He's a big fan of both the iOS and Android platforms and thinks both have their merits. Julian Richards is a freelance writer and gaming aficionado. With 25 years of console gaming under his belt, he's a keen enthusiast of gaming on any platform and has followed the progression of gaming on mobile devices with great interest.

OUR PICKS

Although the chapters do not rank the games featured in order of merit, at the beginning of each we give you our personal Top 10 for that genre. At the end of the book, you will also find four **alternative Top 20 lists** from Flame Tree's point of view, just to provide some extra food for thought. These have been compiled using similar criteria to that mentioned above, and include many games featured or mentioned in this book as well as a few extras.

Join the Debate

What do you think of what we've selected? What have we missed out, got wrong? What would be your Top 10? Take a look at the **Flame Tree Publishing** on **Facebook**.

ARCADE & ADVENTURE GAMES

ARCADE & ADVENTURE GAMES

Arcade and adventure titles are the lifeblood of touchscreen-based mobile gaming. In this section you'll meet some loveable new characters, encounter some seemingly unbeatable foes, enter new worlds and even catch up with some familiar faces.

In this chapter you'll find some of the most popular, unique and intuitive titles available today. From games like the simple, hand-drawn Doodle Jump to the highly sophisticated Infinity Blade, with console-quality visuals and gameplay, there's something for everyone. There's even an acrobatic, roller-skating grandmother.

SIMPLE AND ADDICTIVE

Where better to start than with Angry Birds? Having amassed over a billion downloads worldwide, it's the one mobile game, above all others, you're most likely to have already played. There's also a collection of the dangerously addictive endless-runner games. In Temple Run,

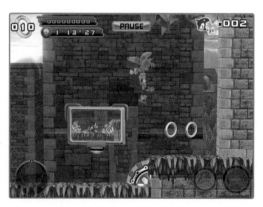

Above: The ever-popular Sonic the Hedgehog has been reinvented for the modern era.

Whale Trail, Ski Safari and Jetpack Joyride, you must survive as long as possible before succumbing to foes and obstacles. Like many games in this section, the beauty lies in their endearing visuals and simple, one-touch gameplay.

Favourites from yesteryear – Rayman and Sonic the Hedgehog – also made the cut after having successfully reinvented themselves for the modern era, while the likes of PewPew and Syder Arcade HD have reimagined classic, space-based shooters for the touchscreen generation. Movie fans will enjoy

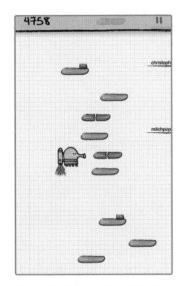

Above: Doodle Jump looks simple, but is highly addictive!

Star Wars Pinball and the excellent Toy Story: Smash It!, in which you get to be Buzz Lightyear.

BATTLE STATIONS!

If you're ready for a battle, we have some brilliant combat games in store for you. Robot-fighting game Epoch is a personal favourite thanks to its intelligent touchscreen-friendly controls and dazzling visuals, while Modern Combat 4 is the closest thing you'll get to a full-on, console-quality, first-person shooter on a mobile device.

LATEST TECHNOLOGY

All of the games in this chapter make the absolute most of the technology available in today's phones and tablets. Games like Temple Run, Air Penguin and Frisbee Forever use the device's gyrometer, allowing you to control your character by tilting the device. Zombies, Run! even uses the GPS functionality on your smartphone to track your progress.

These titles use a combination of touchscreen swipes, slides, taps and more, with intuitive control systems designed specifically for playing on mobile devices. Indeed, the best games in this section – like Tiny Wings and Granny Smith – make you forget you ever used a control pad for gaming. It's the sheer simplicity of these games that makes them so addictive and enjoyable. You'll pick them up in no time, but won't be able to put them down.

Above: In Frisbee Forever you can guide the frisbee by tilting your device.

CHRIS'S PICKS

We hope you enjoy playing these games as much as we did. For the record, here's my personal Arcade and Adventure Top 10:

❶ RAYMAN JUNGLE RUN (see pages 50–51)
Console-favourite Rayman jumps (and runs and flies) on to mobile devices, with a stunning jaunt through the jungle.

❷ WHALE TRAIL (see pages 70–71)
This soothing, psychedelic adventure is the best tap-tap-tap game you'll ever play and the only one to have a catchy, sing-along theme tune.

❸ EPOCH (see pages 32–33)
Take cover and time your attacks in this frenetic, touchscreen-friendly take on the third-person shooter genre.

❹ INFINITY BLADE II (see pages 42–43)
Get medieval on evil warriors and slash your way to vengeance in these incredible fantasy adventures that have shaped the future of mobile gaming.

⑤ TEMPLE RUN *(see pages 64–65)*

A fast-paced, Indiana Jones-inspired, one-slip-and-you're-dead style adventure that'll make you think twice about raiding ancient monuments.

⑥ TINY WINGS *(see pages 66–67)*

Help a cute baby bird overcome his still-developing aviation skills by using the slope of the land.

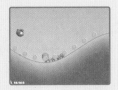

⑦ ANGRY BIRDS *(see pages 26–27)*

You are an Angry Bird, determined to destroy little egg-stealing piggies with your slingshot. If you're new to mobile gaming, this is a pretty good place to start.

⑧ GRANNY SMITH *(see pages 40–41)*

It's broken hips all round as Granny Smith gets her skates on in pursuit of an apple bandit.

⑨ STAR WARS PINBALL *(see pages 56–57)*

This is the pinball game you're looking for. May the force, and the flippers, be with you.

⑩ SYDER ARCADE HD *(see pages 62–63)*

For all-out, white-knuckle action, you really can't beat a space-based shoot 'em up.

AIR PENGUIN

Category: Arcade • **Available on:** Android/iOS • **Cost:** Free/Paid
Download size: Small • **Age rating:** Suitable for all

SYNOPSIS

Help your penguin avoid shark-infested waters and rescue errant family members scattered across the melting glacier.

Bird fever has overtaken iOS and Android devices. You've already read briefly about Angry Birds and Tiny Wings, now the humble, waddling snowbird gets a chance to shine in the adorable and enjoyable Air Penguin. For some reason, this little penguin chooses not to swim. It may be down to his rotund appearance, but those extra couple of pounds around the waist make him very bouncy – hence the premise for this game.

The story requires you to progress through levels to be reunited with your family members, isolated following a glacier melt. To do this, you have to avoid the water by flipping from ice block to ice block. There are creatures that help you along the way, like turtle taxis, ones that get in the way, like seals, and those that just want to eat you – like sharks. A word of warning: don't jump the shark. There's also the survival mode, which challenges you to rack up as many jumps as possible before landing in the drink.

GAMEPLAY

The game mechanics don't require you to touch the screen, but instead rely on your phone or tablet's motion sensors. You flick/tilt the device forwards, backwards and side to side to avoid obstacles and make those all-important landings.

In many ways, playing Air Penguin is more like playing golf. There's a flag that's your target, there are obvious water hazards, bunkers (but with teeth – the open jaws of hungry sharks!) and you need to apply spin to hit the target. The game offers bonus points for completing the level close to the flag. As these levels get maddeningly difficult, you'll probably wish you had a golf club.

THE LOWDOWN

We tried playing Air Penguin on iPad and, although the visuals were beautiful, we found all the tilting and turning a little taxing on the arm muscles, so switched to an iPhone 5. Still, the potential risk of repetitive strain injury is a blessing, otherwise we'd never put Air Penguin down.

You Might Also Like
Doodle Jump (pages 30–31) is a vertical equivalent, while for penguin enthusiasts, Super Penguins is a fun outing.

Hot Tip
Try not to be too jerky. The more smoothly you travel, the easier it is to make adjustments.

ANGRY BIRDS

Category: Arcade • **Available on:** Android/iOS • **Cost:** Free/Paid
Download size: Medium • **Age rating:** Suitable for all

SYNOPSIS

At the last count, 263 million people were playing Angry Birds, in its various guises, every month. If you're new to mobile gaming, this is a pretty good place to start.

Angry Birds is the most popular touchscreen game ever. There's merchandise, cartoons, a movie in production, even a theme park, spawned from over one billion downloads. Business is booming and, in this case, business is flinging cute cartoon birds at pigs. The Angry Birds have every right to be annoyed. They've had their precious eggs stolen by the bacon. It's your job to slingshot through defences erected by the unrepentant armies of sneering green pigs and smash them into oblivion.

GAMEPLAY

The gameplay couldn't be simpler, but mastering the levels and earning all three stars for each gets decidedly tricky and dangerously addictive. Simply hold the screen and pull the catapult

back. The more torque you apply, the greater power. Moving it up and down alters the trajectory. Once you're happy with your aim, let go to send your bird on his way.

If you destroy all pigs with your allotted avian artillery, you'll complete the level, but the real trick is to do so while sacrificing as few birds as possible. Trust us, you'll replay levels over and over to find ways to kill two pigs with one bird.

As the levels get trickier you'll get to use different birds: Red is your stock, well-rounded bird; Blue splits into three when you tap the screen; Yellow can speed up on request; and Black explodes like a bomb when he lands and also when the screen is tapped. You'll soon learn which birds are best for each part of each level.

THE LOWDOWN

There are multiple ways to get your Angry Birds kicks. There are themed efforts like Seasons, Rio, Space (where you'll encounter gravitational problems) and even Star Wars. There are free versions of each, and developer Rovio is always adding new levels.

Some folks never get past Angry Birds in their touchscreen gaming pursuits. The idea of this book is to make sure that isn't the case, but ignoring this game would be like going to Rome without seeing the Colosseum.

You Might Also Like
If you've smashed so many pigs you're starting to feel sorry for them, try Bad Piggies (pages 28–29) and play the game from their point of view.

Hot Tip
If you can't handle the two-star frustration any longer and aren't above cheating, there are walkthroughs on YouTube for pretty much every level.

BAD PIGGIES

Category: Arcade (a bit of puzzle too) • **Available on:** Android/iOS
Cost: Free/Paid • **Download size:** Small • **Age rating:** Suitable for all

SYNOPSIS

So this is why those Angry Birds are so upset! Help the Bad Piggies build complex vehicles and become an accessory to grand theft egg.

After years of playing Angry Birds, we'd come to despise those pesky pigs and their rampant egg thievery, especially their gleeful sneers whenever we failed to conquer their impregnable defences. However, after a few hours playing Bad Piggies and seeing the world from their porcine perspective, we've come to respect just how hard they had to work for those eggs. So, we've called an uneasy truce.

GAMEPLAY

The game starts by detailing the piggies' plan. However, their map to the nest was torn into many pieces, which you now have to recover, and the only way to recover the map is to build Wacky

Races-like contraptions. The components you need – wooden crates, wheels, springs, turbo boosts, engines, propellers, umbrellas, you name it – are all there and you'll need to assemble them by placing them on a grid and setting them on their way. It's trial and error, but you'll learn as you play.

The idea, as in Angry Birds, is to achieve the perfect store of three stars on each level, but this is a lot more challenging than the title that spawned its existence. You'll need to build different configurations for multiple attempts at each level before you achieve perfection. Reaching time goals, opening star crates and completing the level without wrecking your vehicle add up to quite some challenge.

THE LOWDOWN

Bad Piggies may boast the same visuals as Angry Birds and take place in the same universe, but that's where the similarities end. This game could just as easily sit in the Puzzle section (see pages 74–129), as it takes a serious amount of considered thought to build and steer these contraptions to their goals. However, the reward is completely worth it.

There's a free version to cut your teeth on, but the full version brings access to all levels and is a mere 69p/99c.

You Might Also Like
Bad Piggies is more like Rovio's Amazing Alex (pages 80–81) than Angry Birds. Also, give iBlast Moki a try: It requires you to use bombs and build contraptions to move the characters through levels.

Hot Tip
When possible, build longer vehicles. They're more balanced and less liable to tip over on your run.

DOODLE JUMP

Category: Arcade • **Available on:** Android/iOS • **Cost:** Free/Paid
Download size: Small • **Age rating:** Suitable for all

SYNOPSIS

An incredibly enjoyable vertical platform game that's as charming as it is addictive.

Throughout this book, we'll emphasize how simplicity is often the key to a thoroughly enjoyable mobile gaming experience. And, in the nicest possible way, what could be simpler than a game that looks like it's been scribbled in a notebook? Doodle Jump is one of those titles that have become synonymous with mobile gaming due to the sheer unabashed fun it offers. In fact, the challenge isn't so much the game itself, but keeping the goofy smile off your face when playing it.

GAMEPLAY

Your task is to scale a never-ending vertical universe by leaping from one platform to the next and, to get all lyrical on you, lifting your character higher than he's ever been lifted before. The cute protagonist automatically does the jumping, so you'll need to tilt the device from left to right in order to ensure he lands safely on platforms. Sounds easy, right?

Green platforms are your stock landing spots, while you'll need to avoid wooden ones. There are also blue platforms that move horizontally and vertically across the screen. If you're lucky, platforms will be littered with springs, trampolines, propeller hats and rocket packs! If you miss your platform, he's sometimes able to land lower down and try again. However, miss those and it's curtains, a fact illustrated by the patented Wile E. Coyote falling-to-his-doom sound effect.

Adding to the challenge, there are growling, grumpy, ugly monsters you have to shoot down (the only time you need to touch the screen) and black holes in the page you have to avoid. As you progress through the level, there are more obstacles and fewer safe havens.

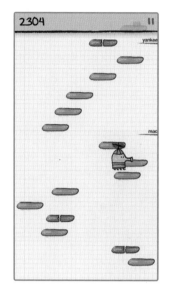 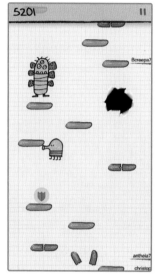

THE LOWDOWN

Scores achieved by other players are listed on the right of the screen and, as you pass them, your own efforts will be added to the leader board. You can also share a particularly fine run on Facebook and Twitter.

Most games feature a 'Play Again' button. We guarantee you'll never use it as much as you will in Doodle Jump.

You Might Also Like
Sonic Jump applies the Doodle Jump concept to the Sonic the Hedgehog universe. Ready to grab some rings?

Hot Tip
Remember when in online multiplayer battles against friends or randoms that it's a battle of attrition, not a race. Stay alive and outlast your foe.

EPOCH

Category: Adventure • **Available on:** Android/iOS • **Cost:** Paid
Download size: Medium • **Age rating:** 12+

SYNOPSIS

Take cover and time your attacks in this frenetic, touchscreen-friendly take on the third-person shooter genre.

There are plenty of excellent shooter games available for iPhone and Android devices, but Epoch stands out from the crowd as perhaps the most enjoyable and definitely the most playable. The action takes place in a post-apocalyptic world in which armies of robots have turned on each other. Some remain loyal to their human masters and others are hellbent on extermination. Your character's mission is to rescue the princess it was assigned to protect. To do so, it must battle endless waves of androids across a ruined cityscape and piece together events by salvaging information from fallen foes.

GAMEPLAY

The key to Epoch's success is its control system. Developer Uppercut Games has realized that a touchscreen is different to a game controller and must be treated as such. So, rather than just adding a bunch of awkward on-screen buttons, it developed its own methods that require a host of intuitive taps and swipes. Tapping a target focuses your fire, while swiping left and right allows you to avoid enemy attacks. Swipe down to take cover and up again to emerge and begin firing. You're constantly on the move, ducking and diving around the screen to avoid attacks and counter with your own. You can also use weapons like grenades, missiles and boosters by tapping the appropriate icons.

This is how touchscreen gaming should be. The action is fast-paced, but the visuals never falter. Indeed, Epoch's gameplay is so breathless that you'll often feel a sense of relief when you've cleared a stage. Doing so unlocks new pieces of information to help you on your way, while you'll also power up your character and unlock new weapons, which can be accessed in the Scrapyard.

THE LOWDOWN

If you feel like a break from the frenetic story mode, you can step into the Arena, which requires you to keep turning robot-killing machines into scrap metal until you finally succumb. It's not for the weak of heart.

All in all, thanks to its innovative touchscreen controls, great visuals and immersive story, this is one of those titles that closes the gap between mobile and console gaming. It's Epoch-al.

You Might Also Like
The Modern Combat (pages 46–47) and Shadowgun games are other fine examples of shooter games for your mobile device.

Hot Tip
When facing multiple enemies, concentrate on the one who can do the most damage. This isn't always the most powerful robot.

FLIGHT CONTROL

Category: Arcade • **Available on:** Android/iOS • **Cost:** Free/Paid
Download size: Small • **Age rating:** Suitable for all

SYNOPSIS

Take a crash course in air-traffic control and make sure everyone gets home safely... well, at least some people. There was a time when we thought it might be fun to be an air-traffic controller. That was before we played Flight Control. Now we just have high blood pressure and a greater respect for the profession.

The idea of this classic mobile game is to land incoming planes on three runways. All you need to do is grab the colour-coded planes, helicopters and jets, and use a finger to draw their path to the appropriate landing zone, all the way to touchdown. It sounds straightforward enough. However, before you know it, there'll be eight planes on the screen, all on a collision course, that need to be directed to three different runways. What do you do? Panic. One crash and it's game over.

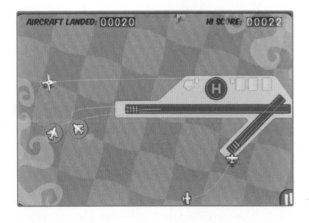

GAMEPLAY

There's a very clear strategy to employ when things get congested. It's important to stagger the approaches and get all of your ducks in a row. The fastest way to the runway is a straight line, but you'll need to give some planes longer, curved approaches to ensure they don't collide on the way in. If things are getting too tight and a collision is imminent, send a plane to circle the runway, just like in real life.

The gameplay gets extremely frenetic and you'll need eyes everywhere until the game inevitably ends with a crash, although you can buy Rewinds to undo your error. The free version of the game is enough to get by on, but the advertisements are quite obtrusive and buying the full version gets rid of these. It also unlocks new maps.

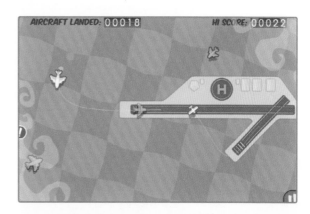

THE LOWDOWN

Flight Control has seemingly been around for ever and is one of the first games we can remember playing on smartphones. The years have been kind to it. The graphics are primitive compared to most titles you'll see within this book, but it's all about that gameplay, which remains as fun and addictive as ever.

You Might Also Like
Tired of sitting in the control tower? Swap it for the cockpit and land some fighter planes Top Gun-style in F18 Carrier Landing.

Hot Tip
Rather than always plotting a route home, sometimes you'll just want to briefly push planes out of the way to deal with more urgent problems.

FRISBEE FOREVER 1 AND 2

Category: Arcade • **Available on:** Android/iOS • **Cost:** Free (with in-app purchases) • **Download size:** Small • **Age rating:** Suitable for all

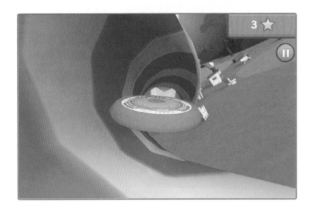

SYNOPSIS

Guide your Frisbee across a host of bright, colourful and challenging obstacle courses in this fun, addictive arcade outing.

Tossing the old Frisbee around is a wonderful pastime. It brings to mind hot summers, days at the beach and being patient with ill-skilled small children. However, the trouble with those flimsy plastic discs is, once they leave your hand, there's very little you can do to control their fate. That's where Frisbee Forever comes in. Here, you're able to 'drive' the Frisbee all the way to its intended target.

The idea is to navigate your way along a host of cartoon-like levels by passing through and avoiding obstacles, keeping up your momentum and making it all the way to the finishing hoop.

GAMEPLAY

To fire off your Frisbee, simply flick the touchscreen. You can then use tilt or tap controls to guide it through the various hoops, loops and corkscrews while steering clear of trees, buildings, cows and so on. Hit an obstacle and your try is over. Miss momentum-maintaining hoops, helpful wind machines and hot routes, and watch as your Frisbee floats to the ground.

The early stages are very easy, but later levels quickly get more challenging, with more difficult obstacles to dodge. Also, you need to do this with enough finesse to earn a gold medal. The stars you obtain along the way are converted into Star Coins, allowing you to unlock more environments and buy new Frisbees.

THE LOWDOWN

Frisbee Forever brings 10 great-looking environments, each containing 10 well-designed levels, meaning – you guessed it – 100 different obstacle courses, plus a host of challenging bonus levels for you to master, while new levels are often added.

Both Frisbee Forever and its sequel are free to download, but you'll have to put in some hours in order to unlock some of the levels. So you may want to spend a little real money on the Star Coins to cut down on the graft necessary to get maximum enjoyment. Indeed, considering it's free, there are few games that offer more entertainment than the Frisbee Forever series. Fire away!

You Might Also Like
If you enjoyed Frisbee Forever, try train dodging in Subway Surfers from the same developer.

Hot Tip
Even if failure is imminent, don't restart the level – you'll still receive all of the stars earned if you see it through.

FRUIT NINJA

Category: Arcade • **Available on:** Android/iOS • **Cost:** Free/Paid
Download size: Small • **Age rating:** Suitable for all

SYNOPSIS

Fed up of your five a day? Use your sweet ninja skills to exact revenge upon them in one of the most popular mobile games ever.

Smartphone gaming hasn't really been around long enough for titles to acquire legendary status, but perhaps 10 years from now, the people who decide these matters will talk about Fruit Ninja in the same way Space Invaders and Super Mario Bros are spoken about today.

GAMEPLAY

The premise of this game couldn't be simpler and, in many ways, that's the beauty of it. Anyone can play, but hardly anyone can stop playing. When a piece of fruit appears, swipe the screen to slice it in two. When more than one piece of fruit jumps up, aim your swipe to take out multiple vitamin-packed items at once. This earns combination points and requires less slicing.

If you haven't sliced the fruit before it drops off the bottom of the screen, you'll lose a life. When bombs appear masquerading as melons, DON'T slice them. They'll end your ninja existence with aplomb. If you perform well in a game, you'll earn star fruit, which can be spent on power-ups like bomb deflectors, while you'll also unlock new blades from the store, for example, when you slice three pineapples in a row.

THE LOWDOWN

There's a free version of Fruit Ninja, which allows you to play the classic mode to your heart's content, but the full game is as cheap as they come and includes an exhilarating online multiplayer mode where you and an opponent are going after the same fruit. Spend a few pennies for the full-fat title and you'll also unlock Zen mode, which has no bombs, unlimited lives and allows you to hone your skills.

Fruit Ninja is one of the few games that's easier to play on a smartphone than a tablet. Granted, the margin of error is smaller on a phone, but when playing on a 10-inch tablet you end up throwing your hands around like an orchestra conductor. You just look a little silly.

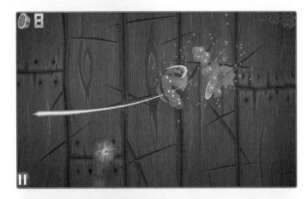

You Might Also Like

There are many Fruit Ninja clones, but none as enjoyable. Instead, extend the fun by downloading the Puss In Boots version.

Hot Tip

Be patient! Slashing one fruit keeps the game ticking over, but taking out multiple items is where the money's at. Before you slice, hang fire for a fraction of a second to see what's coming up next.

GRANNY SMITH

Category: Arcade • **Available on:** Android/iOS • **Cost:** Free/Paid
Download size: Small • **Age rating:** Suitable for all

SYNOPSIS

It's broken hips all round as Granny Smith gets her skates on in pursuit of an apple bandit.

What immediately stands out about Granny Smith – aside from the elderly lady speeding on roller skates, somersaulting and zip-lining across telephone lines – is the pristine, beautiful graphics. The game looks like how you'd imagine a Disney Pixar adaptation of Wallace & Gromit, but it's not all about the design; Granny Smith is also pure unadulterated fun to play.

Your mission is to chase down a thief who stole apples from your garden. You're armed with roller skates, a cane, some surprisingly acrobatic flair and a determination that sees Granny willingly crash through windows and roofs.

GAMEPLAY

There are three apples littered throughout the run, which you must reclaim before the fruit thief gets his grubby paws on them. Jump over small obstacles by tapping the screen and somersault larger ones by holding and releasing to time the landing. You'll earn points and momentum for a good landing. Mistime it and ... splat! Down goes Granny, purse and all. When you see a telephone wire or a swing, somersault towards it and grab it with your cane. Miss and it's a headfirst fall to the floor.

As well as swiping the thief with your cane during the run, you can use coins collected throughout the levels to buy power-ups, such as baseballs to throw, banana skins to leave and a helmet for protection. Once you've completed the level, you can even go back and watch it as a vintage movie, featuring immense slow-motion crashes.

THE LOWDOWN

The early levels are easy to complete and more about attaining the perfect score, but as you move through the worlds (there are four in total) things get decidedly trickier and simply getting through them is enough of a challenge – especially those zero-gravity levels in the space world!

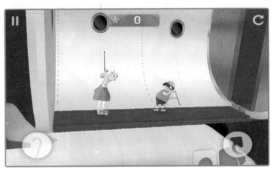

All in all, Granny Smith is one of those few games that ticks all of the boxes. It looks great, plays great, has plenty of replay value and gets an A for entertainment. The free version gets you one world to play in, but the full game is a must.

You Might Also Like
Developer Mediocre AB also built the fun, water-based physics game Sprinkle (pages 112–13). It's definitely worth a try if you enjoyed Granny Smith.

Hot Tip
Crashing through windows is one of the most fun aspects of the game, but it does slow you down. Lob a baseball through it first to maintain your momentum.

INFINITY BLADE AND INFINITY BLADE II

Category: Adventure • **Available on:** iOS • **Cost:** Paid
Download size: Large • **Age rating:** Mature

SYNOPSIS

Get medieval on evil warriors and slash your way to vengeance in these incredible fantasy adventures that have shaped the future of mobile gaming.

The Infinity Blade games have become a standard bearer for iOS gaming. The combination of console-quality visuals, easy-to-master touchscreen battles and in-depth storylines is bested by nothing. Quite simply, you have to try these games.

In the first award-winning title, your warrior's MO is to kill the God King, who murdered your father, and take from him his all-powerful Infinity Blade. In the sequel... well, we don't want to spoil the surprise. In both games, the idea is the same. You'll need to conquer a host of fearsome and often gruesome foes, each tougher than the last, in order to progress towards the final battle.

GAMEPLAY

Once you enter these beautifully rendered battle scenes, the intuitive gameplay is a joy. You'll need to block (hold the screen), dodge (tap left or right) and parry (swipe up) between launching your own attacks. You do this by slashing at the screen in a host of directions in order to cause the maximum damage. But there's more to it than that. You can build up the Super Attack meter in order to inflict more damage, while drawing shapes on the screen at appropriate times launches magic spells.

When you've drained your opponent's energy, you can bring it to an end with a sweet finishing combination and a final death scene that's definitely not suitable for children. After each battle, you'll earn XP (experience points), which must be spent on upgrading your armour, weapons and skills to enable you to match the tougher challenges ahead.

THE LOWDOWN

As we mentioned, the storytelling is beyond anything we've seen on mobile devices; the movie-quality sequences are as entertaining as the gameplay itself and there are plenty of twists. Thanks to Epic Games' Unreal Engine, it's also the best-looking game available on iOS. The level of detail is absolutely breathtaking. However, that does mean quite a large download size.

Unfortunately, Chair Entertainment hasn't launched Infinity Blade on Android, but such is the quality of this outing, it's almost worth buying an iPhone or an iPad!

You Might Also Like
For more console-quality mobile gaming, download the Mass Effect or Dead Space series.

Hot Tip
Mix up your attacks. Swiping left–right–left is the most damage you can do in a three-hit combo. Also, be on the lookout for health vials and cash lying around.

JETPACK JOYRIDE

Category: Arcade • **Available on**: Android/iOS • **Cost**: Free (with in-app purchases) • **Download size**: Small • **Age rating**: Suitable for all

SYNOPSIS

The clue's in the name! You must help all-action hero Barry Steakfries commandeer a host of experimental jetpacks from a top-secret science lab.

Everyone knows that secret laboratories are full of evildoers, plotting the downfall of humanity. So we're glad that men like Barry Steakfries – Jetpack Joyride's hero – exist to ensure they don't get away with it.

In this massively addictive, highly playable game, you start with a machine-gun-powered jetpack, which you use to levitate around the levels, avoiding obstacles like missiles, electric charges and lasers, all of which toast you with a single blow. You can also collect coins and spin tokens, while obtaining awesome vehicles to assist your progress.

GAMEPLAY

The one-touch gameplay is in the same vein as Whale Trail (see pages 70–71). You must tap the screen to power the jetpack and hold it to stay in the air. Rapid taps keep you stable, while letting go brings Barry down to the floor. The real fun starts when you pick up vehicles like the Bad As Hog motorcycle, a fire-breathing dragon called Mr Cuddles, a teleporter that helps Barry avoid trouble, or even an antigravity suit! They all act and are controlled differently.

If you've picked up Spin Tokens along the way, you'll get the chance to earn bonuses once you eventually succumb by spinning the fruit-machine reels. These can gain you extra lives, power-ups, coins for your Stash or distance bonuses to boost your score. As well as posting distance scores, you can knock off the in-game Missions – for example, travel 3,000 metres or use two vehicles in one run.

THE LOWDOWN

Amazingly, Jetpack Joyride is a free game, yet it isn't one that requires you to spend to get full enjoyment. Better jetpacks, power-ups and costumes can be bought with real money, but that's more of a shortcut than a necessity. Through playing the game and collecting coins, everything can be earned.

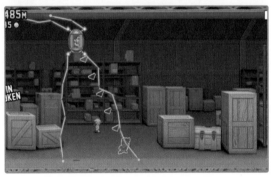

We loved everything about Jetpack Joyride, from its 16-bit-inspired visuals, loveable central character and addictive one-touch controls to its catchy soundtrack. No mobile gaming collection is complete without it.

You Might Also Like
Fallen in love with Barry Steakfries? His adventures continue in Monster Dash.

Hot Tip
Make the first Gadget you buy with your Stash, the Air Barrys trainers. They make it easier to jump the electrical charges.

MODERN COMBAT 4: ZERO HOUR

Category: Arcade • **Available on:** Android/iOS • **Cost:** Paid
Download size: Large • **Age rating:** Mature

SYNOPSIS

The best first-person shooter game available for mobile devices. As the game's voiceover proclaims: 'This is a new age of combat.'

The first-person shooter – where the player sees the battle through the eyes of the soldier – is the most popular games genre on Earth, but until Gameloft's Modern Combat series came along, there wasn't really a blockbusting mobile equivalent. Zero Hour, which is the fourth instalment, is the best, most detailed and most enjoyable yet. The scenarios presented by this type of game, however, are never sunshine and rainbows. This time, your elite team is charged with rescuing global leaders following synchronized nuclear attacks across the globe.

GAMEPLAY

You control your character using various gestures and customizable on-screen buttons with aiming, firing, crouching and reloading in easy-to-reach places. You look around the landscape by swiping the screen, but there's also a sweet gyroscope mode, which involves turning the device to change the view.

In the campaign mode, you have to carry out various missions, on foot, from vehicles and sometimes from the air, while the battlefields vary from sunny Barcelona all the way to Antarctica.

At times, the story also requires you to play as the evil terrorist who masterminded the attack.

THE LOWDOWN

The story mode is vast and fantastic, but it's the free online multiplayer mode that'll keep you coming back. Here you can battle or join forces with thousands of other players in order to complete missions. Alternatively, you can simply vie to win an all-out death match and move up the leader boards.

The game itself is massive and takes an age to download, but as soon as we saw the sheer scale and the incredible console-quality visuals, we stopped complaining. At around a fiver, Modern Combat 4 is also one of the more expensive mobile games in this book,

but you'll see where the extra couple of quid went. When you consider it's around a tenth of the cost of the latest Call of Duty console game, it's not bad value at all.

You Might Also Like
Call of Duty: World at War Zombies is also available for mobile, but iOS owners can also take a trip down Memory Lane with the original DOOM games.

Hot Tip
The Aim Assist is a useful tool. When you tap the aim button, it'll automatically focus on to your target, allowing you to take him out more easily.

PEWPEW

Category: Arcade • **Available on:** Android/iOS • **Cost:** Free
Download size: Small • **Age rating:** 12+

SYNOPSIS

Don't be put off by the retro visuals. It's the retro gameplay that'll keep you coming back for more.

PewPew is great little time-killer, inspired by Atari's Asteroids. In its heyday, Asteroids was the most popular game on the planet, but that was over three decades ago and what was once played on a massive arcade machine now fits in the palm of our hands. Naturally, the homage isn't quite as primitive as the dodge-and-shoot classic, but the spirit is certainly felt. The idea is to guide a little spaceship around a gaming grid, firing on constantly spawning enemies of differing neon colours, shapes and sizes. They don't fire back, but making contact with them damages your ship and eventually kills you. Attack is the best form of defence, but you can take refuge behind items on the board, while also collecting power-ups to boost your ship's firepower and shield.

GAMEPLAY

You move the spaceship around the grid by holding down and rotating a yellow directional button. Firing is controlled by the adjacent red button, which can be rotated to control the direction of the fire. Once you get used to it, the two-finger combination that enables you to weave through ever-increasing numbers of enemies, or blast them out of your way, or duck into pick up power-ups, is really intuitive and lots of fun.

In all modes of this fast-paced arcade game, the idea is simply to survive as long as possible. It gets quite hard quite quickly, so expect to play in multiple short bursts, rather than embarking on an epic space mission. Do well and you'll earn medals that can be used to unlock new ships.

THE LOWDOWN

Beyond the basic Pandemonium mode, there are plenty of variations to keep things interesting. The Dodge mode requires no firepower, just careful manoeuvring, while Chromatic Combat requires you to kill enemies the same colour as your ship. There's also an Asteroids mode for the nostalgic.

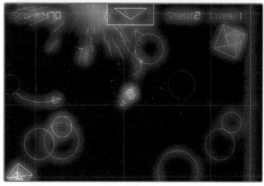

Mobile games that give new life to retro classics are always welcomed and the completely free PewPew is, as its name suggests, a blast.

You Might Also Like
Download the original Asteroids through the Atari Greatest Hits app. Asteroids Gunner is also worth a try. If you can't get enough PewPew, the sequel is now available.

Hot Tip
The arrows that sometimes appear next to your ship guide you to power-up boxes.

RAYMAN JUNGLE RUN

Category: Arcade • **Available on:** Android/iOS • **Cost:** Paid
Download size: Small • **Age rating:** Suitable for all

SYNOPSIS

Console favourite Rayman jumps (and runs and flies) on to mobile devices with a stunning jaunt through the jungle.

This exquisite, hugely detailed, console-quality adventure is a platform game from the old school. It's simple to pick up, fun to play and practically impossible to put down. Indeed, the game looks so good, plays so well and has such depth that it wouldn't look out of place on an Xbox or PlayStation console.

GAMEPLAY

In Jungle Run, our hero loves to run and, as soon as you press Go, he doesn't stop. However, to progress through the countless levels within multiple themed worlds, he needs to jump, fly and punch, and that's where you come in. You travel through the well-thought-out levels, swinging through trees, walking up walls, flying over obstacles and much more.

The early levels aren't that hard to complete, but the idea is to progress through each with a perfect score. To do that, you'll need to grab all of the Lums (the weird little rabbit-like creatures) and coins littered around the levels. Trust us, you'll keep playing until you catch them all, because a perfect score nets you a ruby at the end of the level. Collect enough rubies to replace Mister Death's missing teeth and he'll kindly unlock the Land of the Dead levels. However, you may not thank him. They're littered with spikes and deadly drops and they're much, much harder. You'll have to play many times before mastering them. At least we did.

THE LOWDOWN

We fell in love with Jungle Run before even starting to play. Once you open the app, the mandolin-strumming central character serenades you and the feel-good fun starts and just doesn't end. The visuals are among the best you'll ever see on a mobile game, the touchscreen gameplay is extremely

well thought out and, at less than the price of a loaf of bread, it offers stunning value. Don't believe us? Load the App Store page for this game and you'll see something quite rare: a full five-star rating, from over 8,000 reviews.

You Might Also Like
In terms of quality, Rayman Jungle Run is peerless. For more of the same, try the Sonic games.

Hot Tip
Collect the hearts littered around the levels. They'll allow you to withstand one hit from enemies.

SKI SAFARI

Category: Arcade • **Available on:** Android/iOS • **Cost:** Paid
Download size: Medium • **Age rating:** Suitable for all

SYNOPSIS

An avalanche is coming to bury you alive, but luckily you have more than a pair of skis to help
you escape.

The last thing anyone needs, when they're snuggled up in their log cabin, is a yeti yelling 'Avalanche!'
at the top of his lungs. But such is life. These things are sent to challenge us. In Ski Safari you're
immediately confronted with the horror of your impending doom as the avalanche sweeps through
your cabin and casts you on to the slopes, duvet 'n' all. Luckily, you slept in your skis.

GAMEPLAY

The challenge is to stay ahead of the deadly snowy apocalypse by jumping obstacles and
performing flips by tapping and holding the screen with a single finger. If you wipe out, the
avalanche draws closer. If that happens, tap the screen rapidly in order to quickly get up. You

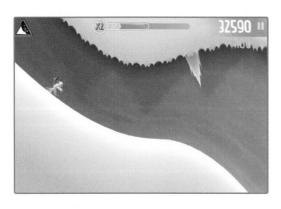

pick up speed by collecting coins and avoiding
collisions, but also by latching on to penguins
and using them as skis, piggybacking on yetis,
and soaring on the wings of eagles. There's even
rocket-powered snowmobiles you can hijack, as
well as wild wolves and reindeer to aid you.

Big points tallies come as you build the
combination meter, thanks to bonuses like
catching penguins in midair, sliding across

clouds and stringing tricks together. Crashing through rocks and ice or mistiming landings resets the multiplier to zero, so it's important to keep your run tidy.

THE LOWDOWN

Ski Safari is extremely simple to pick up. It looks absolutely fabulous and the in-game challenges (such as landing a double backflip, catching five penguins in the air) keep things interesting by unlocking new items and adding multipliers to improve your scores. There's also a store where you can spend coins on new characters, courses and bonuses. The game is more rewarding the more you play.

The endless-run style is in the same school as Temple Run (see pages 64–65). Unfortunately, regardless of how well you do, you will meet your end eventually. Still, there are worse ways to go than backflipping on skis, speeding on snowmobiles, flying with eagles and surfing on yetis.

You Might Also Like

Defiant Development, the makers of Ski Safari, also brought us the equally awesome Rocket Bunnies. Talented folks.

Hot Tip

Catch an eagle during a jump and they'll carry you far from danger. When they drop you, immediately hold the screen to perform a quadruple backflip.

SONIC THE HEDGEHOG 4 (EPISODES 1 AND 2)

Category: Arcade • **Available on**: Android/iOS • **Cost**: Free/Paid
Download size: Medium • **Age rating**: Suitable for all

SYNOPSIS

Twenty years after his console heyday, Sonic is back with an episodic new adventure for iOS and Android devices.

Sonic the Hedgehog is gaming royalty, so it seems only right that one of the most celebrated characters of all has a place in this new touchscreen era – especially when his old rival Mario is nowhere to be seen. The new adventure, 16 years in the making, sees Sonic return to his 2D platform roots, spinning his way around intricate levels, looping the loop, collecting rings, smiting bad robots and chasing down the evil Dr Eggman, who, once again, has kidnapped Sonic's animal pals, and that's whom you're rescuing.

Fans of the original games will be pleased to hear that SEGA hasn't messed with the formula too much, but for those who've never played a Sonic game before, there's plenty to get you hooked.

The level designs are quite familiar (Splash Hill and Casino Street in Episode 1 (especially), along with the classic theme songs, sound effects and power-ups, like invincibility and speed shoes, which can be obtained by smashing the TV sets littered around the levels.

GAMEPLAY

The gameplay is also very similar. Move Sonic with the directional pad (or D-pad for short), hit the button to jump and use the two together to spin and perform targeted attacks and jumps with more momentum. If you get hit, you lose your rings. If you're hit while ringless, you lose a life. Collect 100 rings and you gain an extra life. In each game, you'll play through four zones, each of which has three acts and a boss level at the end. Here, you'll face off against Dr Eggman in one of his evil contraptions.

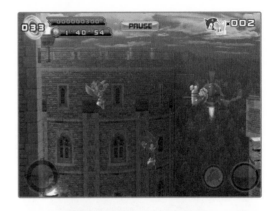

THE LOWDOWN

Episode 1 is a fun throwback, but Episode 2 is a brilliant game in its own right. The visuals are absolutely superb, the gameplay is faster, more immersive, better thought out and, unlike the first game, it feels like it belongs on a touchscreen. If you're only going to download one of the two, make sure it's this one! We can't wait for Episode 3.

You Might Also Like
Rayman Jungle Run (pages 50–51) is the best game of its kind for mobile devices, but if you're all about Sonic, there are tons of his adventures out there.

Hot Tip
Complete a level with 50 rings or more and you'll see a huge gold ring. Jump through it and play the bonus level to obtain Chaos Emeralds.

STAR WARS PINBALL

Category: Arcade • **Available on:** Android/iOS • **Cost:** Paid
Download size: Small • **Age rating:** Suitable for all

SYNOPSIS

This is the pinball game you're looking for. May the force, and the flippers, be with you.

Adding a Star Wars logo always gets more eyes on a product, but this is one of the few occasions when the game lives up to the name. A huge amount of thought has gone into making Star Wars Pinball a dream to play, while offering a gaming experience even non-Jedi wannabes will struggle to put down.

The idea is to send your ball flying across the intricately themed tables, unlocking various scenes from the movie, and then completing the various multifaceted challenges. For example, hitting the right lane a number of times helps Han Solo land the Millennium Falcon on Cloud City. Precise details on how to win the challenges can be found in the table guide screen when pausing the action. At various points during the challenges, you'll even see animated characters like Luke Skywalker crawl out on to the playing surface. Dare to hit Darth Vader when he appears and he'll crush your pinball like a Malteser!

GAMEPLAY

The basic gameplay is as you'd expect for a pinball simulator. Tap and hold the screen to activate the flippers. The tables look fantastic and are hugely detailed and, to that end, we preferred playing on a larger tablet screen. However, it's the little touches that'll really put a smile on your face, like using a lightsaber to push the ball into play, or the aptly timed quotes and sound-bites from the movie. Heck, even the agony of losing a ball is tempered by Han Solo proclaiming: 'You said you wanted to be around when I made a mistake, well this could be it sweetheart.'

Another fun element allows you to choose whether you want to represent the Light Side or the Dark Side, with all scores contributing to a global online tally!

THE LOWDOWN

The game comes with an Empire Strikes Back-themed table, while Clone Wars and Bobba Fett tables are available as in-app purchases for the same price. More tables are coming soon, but in the meantime there's so much to discover and play for in a galaxy far, far away.

You Might Also Like
Initially, we had the excellent War Pinball shortlisted for this book, but what can we say? We're suckers for Star Wars.

Hot Tip
Light up the Training indicator to unlock a neat mini-game where Luke has to fend off droid blasts with his lightsaber.

SUPER HEXAGON

Category: Arcade • **Available on:** Android/iOS • **Cost:** Paid
Download size: Small • **Age rating:** Suitable for all

SYNOPSIS

Ready for a challenge? Because this transfixing, fast-paced adventure through geometry is perhaps the hardest game you'll ever play.

Super Hexagon is an extremely neat little game, but mightily tricky to master. You control a tiny triangle and have to weave it through small gaps in an endlessly shifting, kaleidoscopic maze of spinning, pulsating and direction-changing shapes. The idea is simple: to survive as long as possible by avoiding any sort of contact with the shapes. You do this by swinging your triangle left and right around the central hexagon. In principle, the idea is to collect enough 'lines' to form a Super Hexagon and complete the level. There are six to work through.

GAMEPLAY

Inspired by some of the simple arcade games of yesteryear, there's no doubting it's an absolutely spectacular game, with great visuals and a catchy techno soundtrack, but it's also incredibly infuriating. Here's how our first 150 attempts to play went in real time. Start. Game over. Again. Game over. Again. Game over. Again…. After about an hour, our longest survival time was 9.22 seconds, so we went in search of an 'easy' mode. It turned out we were already in it. It's called Hard. The other options were Harder and Hardest. Great.

So we persevered and so should you. Eventually, we were able to survive for 24 seconds, but death was always just around the corner, usually when the ever-shifting landscape, without warning, turned on its axis from clockwise to anticlockwise.

THE LOWDOWN

Not that we're experts, but the key, it seems, is to make decisions early and stick to them. Identify the gap in the next shape as soon as possible and go for it. There's no time to think, no margin for error and no Matrix moment where you crack the code and suddenly become awesome. If you make it to 30 seconds, you're officially better than us.

So why are we recommending this game? Because despite our endless epic fails, we cannot fault it. We're still playing, hoping beyond hope that next time we'll unlock Super Hexagon and get some peace of mind. Download at your peril.

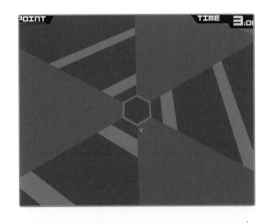

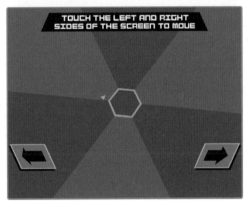

You Might Also Like
The 'Seven Nation Army' music video by The White Stripes. It's not a game, but visually it's just about the only thing on Earth that's as mesmerizing.

Hot Tip
Try to make the larger rotations in one press-and-hold of the screen rather than tapping. Also, try to memorize the patterns, they're repeated throughout the level.

SUPER MONKEY BALL 2

Category: Adventure/Arcade • **Available on:** Android/iOS
Cost: Paid • **Download size:** Small • **Age rating:** Suitable for all

SYNOPSIS

Monkeys! Balls! Need we say more?

Ah monkeys, you're pretty much guaranteed comedy when you put a monkey in nearly any scenario. They're lovable little scamps who get up to all sorts of mischief. However, putting them inside plastic balls? That's a new one on us.

If you're not familiar with the Super Monkey Ball series, we'll give you a brief rundown. It started life as an arcade game (complete with banana-shaped joystick, naturally) before making a name for itself on Nintendo's Gamecube system and subsequently the Playstation and Xbox.

GAMEPLAY

The premise of the game is that you control a monkey, in a ball, and have to guide him to the end of the level, without falling off increasingly difficult tracks, or running out of time. Gameplay consists of you tilting the screen forwards or backwards to move your monkey in the direction you want to go. If you need to turn left or right, then all you have to do is tilt your device left or right. It really is that simple. At first you might struggle with the system, but once you've played a few levels, it becomes second nature and you'll be monkey balling like a pro.

There is a total of five worlds and each world has two sets of 10 levels. Completing one unlocks the next and all the levels are rendered in beautiful 3D. There are bananas dotted around each level and 10 bananas earn you an extra life. Getting them all without falling off the level means you win a crown for that level. If you can get all the crowns in a set, you'll unlock a special bonus crown level.

THE LOWDOWN

The level design in the latter stages is particularly fiendish, so the game is a more thanrespectable challenge. There are also mini games of monkey bowling, golf and target (we were bitterly disappointed that there was no monkey tennis) should you fancy a change of pace.

The game looks and sounds like you would expect from SEGA. With bright, brash colour schemes and bouncy music coupled with over-enthusiastic voice samples. Super Monkey Ball 2 is terrific value for money and it will keep you occupied for ages.

You Might Also Like
Super Monkey Ball 2: Sakura Edition has the monkeys rolling round Far Eastern locations.

Hot Tip
When you're on train tracks, try to stay in the middle; the outside of the tracks will act as barriers to keep you in place.

SYDER ARCADE HD

Category: Arcade • **Available on:** Android/iOS • **Cost:** Paid
Download size: Medium • **Age rating:** 12+ (some moderately bad language)

SYNOPSIS

For all-out, white-knuckle action, you really can't beat a space-based shoot 'em up.

Syder Arcade takes a beloved formula and updates it for the touchscreen era, ramping up the visuals and adding some neat gameplay nuances. Fans of old-school shooters like R-Type (as well as fans of the *Battlestar Galactica* TV series!) will love this game. You're the solo pilot of a ship charged with taking down 'a motherload of alien invaders' advancing on your position. The first thing you'll notice are the gorgeous colourful visuals, but soon you'll be utterly absorbed in the gameplay.

GAMEPLAY

Holding your finger on the touchscreen moves your ship around the gaming area, enabling you to be nimble, evade enemy fire and position yourself perfectly to take down the enemies. You'll fire your basic guns whenever on the move. Some enemies can withstand more firepower than

others and this is where the special weapon comes in. It can be charged and used sporadically to cause huge damage and take out multiple opponents.

While most games of this ilk require users to shoot from left to right, in Syder Arcade HD enemies are advancing on you from both directions, requiring you to flip the ship. You'll need eyes in the back of your head, but thankfully there's a map that helps you keep track of approaching invaders. Your ship – there are three instantly playable vessels to choose from – can sustain some damage, but can be repaired and upgraded with speed and weapons by picking up onscreen power-ups. These are essential for keeping you alive.

THE LOWDOWN

In the campaign mode, there are six missions, of increasing difficulty to complete and also the survival mode, which requires you to keep taking down opponents. To complete missions, you'll need to guide your way through asteroid fields and protect your home ship before gearing up for an almighty boss battle. None of them is easy, but ultimately it's really satisfying when you take the big guy down.

The developers list the essential features as 'shooting' and 'dying', which sums up Syder Arcade HD. It harks back to a simpler gaming age.

You Might Also Like
Syder Arcade HD is very much a throwback to 1990s games like R-Type, which is also available for Android gamers.

Hot Tip
Check out the map. If no more enemies are approaching from one side of the screen, head there and meet all attackers head on.

TEMPLE RUN

Category: Arcade • **Available on:** Android/iOS • **Cost:** Free
Download size: Medium • **Age rating:** Suitable for all

SYNOPSIS

A fast-paced, Indiana Jones-inspired, one-slip-and-you're-dead style adventure that'll make you think twice about raiding ancient monuments.

You're going to meet your end playing Temple Run. There's literally no escape, so you may as well accept your fate. All you can do is prolong the life of doomed tomb-raiding Guy Dangerous long enough to set a new high score.

You join the game with the heist already completed and the character's ill-fated attempts to escape with the loot just begun. Behind you are hideous, cannibalistic monkey-things, waiting to pounce on any mistake, while fire and deadly drops lie in front of you. Not an ideal scenario, really.

GAMEPLAY

The controls, as with all great mobile games, are relatively straightforward. Swipe up and down to jump over or slide under obstacles; swipe left or right to turn corners and tilt left and right to steer your character away from danger. Mistime a jump, and you're dead. Splat. Mistime a slide under fire and you'll be incinerated. Lovely. As the level progresses, things get faster and faster, the jumps start to look like turns and this is when instinct takes over. You'll want to gnaw your own fist off when you hit jump instead of turn and end up as crocodile lunch, but the adrenaline rush is part of what will bring you back to Temple Run over and over again. It looks great, plays great and, quite frankly, is great.

Thankfully, there's respite en route to your doom. As you progress, you can collect gold coins littered around the place. At the end of each turn, these can be spent on power-ups. There are also objectives to unlock, which, along with the humorous messages that accompany your final score, make the endless cycle of death a little less deflating.

THE LOWDOWN

The game's developer Imangi Studios recently launched a sequel that's also free to play. It brings new gameplay elements (ziplines! minecarts!), a visual overhaul and new power-ups. Such was the popularity of the original that the follow-up was downloaded 50 million times in the first two weeks. We actually prefer the simplicity of the original, but Temple Run 2 certainly takes the action to the next level.

You Might Also Like
Agent Dash and Pitfall are variations on the run-run-run-die formula, but Temple Run and its sequel are the best.

Hot Tip
The best power-up is the Magnet. It draws all coins to you without having to worry about steering.

TINY WINGS

Category: Arcade • **Available on:** iPhone • **Cost:** Paid
Download size: Medium • **Age rating:** Suitable for all

SYNOPSIS

That's not flying, it's falling with style! Help a cute baby bird overcome his still-developing aviation skills by using the slope of the land.

Angry Birds isn't the only must-own, bird-centric physics game on the App Store. There's another winged wonder in town. Although it has lived somewhat in the shadow of Angry Birds, Tiny Wings is an entirely different proposition and a truly great mobile game in its own right. The premise of the game is to help a flightless bird to get across a range of islands, making it as far as you can before nightfall.

GAMEPLAY

Tiny Wings can slowly flap his way to the top of hills, but it's up to you to ensure he makes the most of the downward slopes. Holding a finger down on the screen stops the flapping and

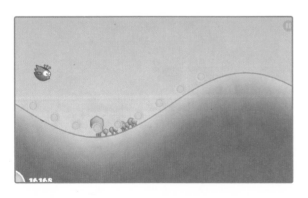

makes him a dense, aerodynamic speedball. You need to let go when he reaches the foot of the upslope, allowing him to spring up the hill and get some air!

His lack of skill in the air means he can't stay up there for long, so you must bring him down at the optimum time. The landing and relaunching process is all about timing and judgement. The more perfect slopes you hit in a row, the greater speed you'll pick up. Hit three in a row and you'll go into Fever mode!

The reward for fast progress is the chance to charter more islands, earn more points and unlock more achievements, and all before nightfall. If the classic mode's not enough fun for you, there's also a neat new Flight School chapter, where you race other birds back to the nest in order to be fed first by mother bird. The iPad version also offers split-screen multiplayer games.

THE LOWDOWN

Once again, the joy in the gameplay is in its one-touch simplicity and it's easy to lose yourself in the smooth, soothing soundtrack and the beautifully drawn visuals.

Tiny Wings is currently only available for iOS devices, so no joy for Android users yet, but in a book called *100 Top Games Apps*, there's no way we could leave it out. It's a truly masterful creation.

You Might Also Like
Those with Android phones should try Tiny Comet and A Tiny Bird's Journey. They're more original than some of the blatant Tiny Wings copies.

Hot Tip
Unlocking achievements in the game allows you to 'Nest Up', which means you can earn more points for your slides.

TOY STORY: SMASH IT!

Category: Adventure (and a bit of puzzle) • **Available on:** Android/iOS
Cost: Paid • **Download size:** Small • **Age rating:** Suitable for all

SYNOPSIS

Utilize the excellent throwing arm of Sir Buzz of Lightyear to take down the minions of Zurg and other baddies in this excellent 3D physics game.

There aren't too many movie spin-off games worth writing home about, but Disney's enjoyable action/puzzler really breaks the mould. This game channels the playful, childlike spirit of the much-beloved films and is one that kids and grown-ups can enjoy in equal measure.

GAMEPLAY

As in the movies, Andy has set up his favourite toys to act out elaborate scenes (such as alien invasions and bank robberies) and it's up to legendary space ranger Buzz Lightyear to save the day. Armed with a host of weapons and an arm like a cannon, Buzz must smash down the structures harbouring those loveable little green alien toys. The aliens themselves can shield the

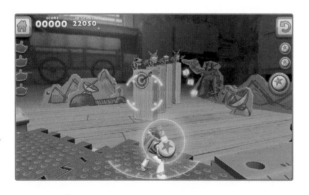

throws, but the building blocks beneath them can be taken down with various weapons. To do this, Buzz has beach balls, rockets and balloons at his disposal. To unleash them, simply drag down on the screen to pull Buzz's arm back, take aim by moving around the touchscreen and then release. Rockets can be made 'unstoppable' after release by pushing the on-screen indicator, while balloons can also be inflated to do more damage.

Buzz can run around the well-crafted levels to find the best line of attack and there are also exploding presents that can be used to take down the aliens, meaning you don't always have to aim directly at them. Force all of the aliens to retreat and you'll complete the level, but to receive a perfect three-star score you'll need to collect the gold blocks littered around the levels (they spell out T-O-Y) and get the job done in the least number of throws possible.

THE LOWDOWN

As with Bad Piggies (see pages 28–29), this game blurs the boundaries between puzzle and action. You'll need to carefully think through your line of attack in order to get

the best scores, while you'll also need a good aim and excellent timing. One well-aimed, well-timed throw from the perfect spot can start a chain reaction that takes out the entire level.

Overall, it's probably a little complicated for younger *Toy Story* fans, but it's a great game to play with the kids, especially if everyone's gathered around a tablet.

You Might Also Like
If you feel like reliving more Disney movie moments, try Nemo's Reef and Wreck-it Ralph.

Hot Tip
When throwing balloons, follow the shadow beneath them to pinpoint the best place to inflate.

WHALE TRAIL

Category: Arcade • **Available on:** Android/iOS • **Cost:** Free/Paid
Download size: Medium • **Age rating:** Suitable for all

SYNOPSIS

This soothing, psychedelic adventure is the best tap-tap-tap game you'll ever play and the only one to have a catchy, sing-along theme tune.

Don't ask us how he got there, because it may have involved mind-altering substances, but Willow the whale is swimming in the clouds and he seems to be having a pretty good time of it. However, as the proverbial fish-out-of-water (well, mammal to be precise), he needs air to survive. As Willow's designated driver, you have to tap or hold the screen to keep him buoyant on his journey across this colourful and imaginative landscape. Holding the screen will allow Willow to fly upwards (keep holding and he loops the loop!), while releasing sees him fall to Earth.

GAMEPLAY

The idea is to follow the Whale Trail of bubbles to keep Willow's air supply up, which builds and multiplies your score. You'll also need to guide him towards the stars littered across the landscape. If you go too long without topping up on bubbles, Willow will fall to his doom at the hands of an evil spider-like cloud monster called Barron Von Barry. If you collide with clouds too often, you're also doomed. Luckily, you have weapons of your own. Collect five stars and you'll send Willow into a Frenzy that makes him temporarily indestructible and able to destroy the grumpy clouds.

During the trail, you'll earn Krill, the in-game currency, which is spent on upgrading Willow's skills. You can make Frenzies last longer and cloud collisions less damaging and also buy treats like Respawn and Instafrenzy to use in the game. You can also unlock Willow's friends as playable characters. These include Gruff from the Super Furry Animals, who sings the dreamy theme song that's as catchy as the game itself.

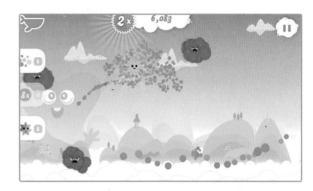

THE LOWDOWN

If you tire of the swim-till-you-drop classic version, there are 96 mini-challenge levels too. The fun really is limitless. It's a simple-to-play, beautiful game that defines everything that a touchscreen title should be.

When compiling the games for this book, we tried to be objective; consider variety, popularity and critical acclaim, but also our personal enjoyment. Whale Trail ticks all these boxes, but mostly the last one. We hope you enjoy it as much as we do.

You Might Also Like
Young kids love Whale Trail, but generally they're terrible at it. Whale Trail Junior is great, but available for iOS only.

Hot Tip
Taking out huge numbers of dark clouds sends your score through the roof, so time your Instafrenzy power-ups carefully.

ZOMBIES, RUN!

Category: Adventure • **Available on:** Android/iOS • **Cost:** Paid
Download size: Medium • **Age rating:** 12+

SYNOPSIS

Gaming isn't really known for its health benefits, but with the moans of the walking dead in your ears, you'll be annihilating your personal bests in no time.

There are some people who really enjoy the challenge of running. For everyone else, there's Zombies, Run! This unique game combines the functionality of a fitness application like Nike+ Running, with a playable element that places you at the centre of a zombie outbreak.

GAMEPLAY

You're Runner 5 (don't ask what happened to the other four) and after crash-landing in the danger zone, you'll be given voice commands to pick up items and supplies to aid fellow survivors and bring them back to base. However, to hang on to those supplies, you'll need to evade the zombies hounding your every step.

When the undead are closing in, you'll hear their groans in your earphones (don't forget them or you can't play) and you must outrun them. The game uses GPS technology and knows how fast you're running, so to avoid capture you need to pick up the pace. You get a 100-metre head start and must stay in front of them.

The stories are well put together and totally engrossing. You'll want to run longer in order to complete the missions and Sam, the radio communications guy who gives you instructions, is hilarious. Because people like to listen to their own music when they run, the commands are interspersed among your own preselected playlists.

THE LOWDOWN

There are over 30 different missions to complete, each lasting around half an hour and each with different goals in mind. At the end of each run, you'll see your time and distance and can share your runs with other survivors at zombiesrungame.com.

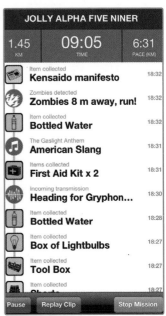

You'll also distribute the supplies you've gained and build the survivor population of your township, which makes it all the more important not to get caught by the zombies and lose your bandages and pain meds. You can do this while you're catching your breath at

You Might Also Like
If you're a running novice, download the Zombies, Run! 5K training app for an eight-week programme to get you through your first three miles.

Hot Tip
If you're a little more advanced in your running, enter the settings and increase the mission length to an hour.

PUZZLE & TRIVIA GAMES

PUZZLE & TRIVIA GAMES

If there's any one category that's suited to modern mobile devices more than any other, then it has to be puzzle and trivia games. Games that were previously played on boards, in magazines and newspapers are now available to take with you wherever you go.

Where once you might have had to survive on one puzzle a day in your newspaper, or had to scribble all over the page to work out your answers, now the choice and range of games are vast and everything is cleaner and much more user friendly.

OLD FAVOURITES WITH A TWIST

So what awaits to pick your brains in this chapter? First we have old favourites reinvented. Board games that you've known since childhood are in here. They've been reimagined and spruced up so that their very best features are emphasized but they're just as you remember them, only brighter and shinier. They've also been created with multiple players in mind and you can now play these classics in any scenario without losing any of the pieces. If that isn't a good argument for the move to digital media, then we don't know what is.

Above: The card game Spider Solitaire is an old favourite, now available to play on the move.

If you like card games, we've sought out the very best available and brought them to you. There is a bewildering array out there when you start to look, so we've taken the hassle out of that for you.

PLAY WITH FRIENDS

We also have the latest in social media games. These are games that encourage you to interact with friends and strangers, actively enhancing the gaming experience. You'll find that these games have new twists on old formats, and we're sure you'll enjoy them as much as we did.

Aside from the old, we also have the new. By that, we mean games that take risks with experimental art styles and bold musical choices, all wrapped around insanely addictive gameplay. From the bizarre of Super Monsters Ate My Condo to the beauty of Contre Jour, these are games that will leave you dazzled by their diversity. They're a break from normal, traditional-style games, but the effort and care that's gone into them make them new favourites.

Above: Contre Jour is a beautiful game that will take you into five strange and abstract worlds.

GOOD FOR YOUR BRAIN!

All of the games featured in this section are amongst the best available to you, and we've selected as broad a range as possible to cater to everyone's tastes, no matter your age. One thing is guaranteed from the games in this chapter: they will all test and tease your brain and keep you absorbed and entertained for hours. Apart from maybe the Moron Test – that one will just make you question your own sanity!

Above: The Moron Test is a challenging brain teaser that will certainly keep you on your toes.

JULIAN'S PICKS

We hope you enjoy playing these games as much as we did. For the record, here's my personal Puzzle and Trivia Top 10:

❶ CONTRE JOUR (*see* pages 84–85)
Guide little Petit through five strange and abstract worlds by manipulating the beautiful, minimalistic landscape.

❷ WORLD OF GOO (*see* pages 126–27)
Video games and art combine in a witty puzzler that will absorb your time like a big ball of goo.

❸ AMAZING ALEX (*see* pages 80–81)
Alex is bored, so what better way to dispel boredom than with incredible physics-based puzzles?

❹ CUT THE ROPE (*see* pages 86–87)
Feed candy to hungry monster Om Nom in this ingenious physics-based puzzler with dangerous levels of cuteness.

❺ THE ROOM (see pages 98–99)

Turn off the lights, turn up the volume and prepare to dedicate your very existence to unravelling the engrossing mystery that is The Room.

❻ TETRIS (see pages 118–19)

One of the most famous video games of all time returns to devour your time once again.

❼ PUDDING MONSTERS HD
(see pages 96–97)

The pudding monsters don't want to get eaten and the only way they can stop that is if they join together. Literally!

❽ SPIRITS (see pages 110–11)

Guide your little white spirits safely to the exit without letting any harm come to them.

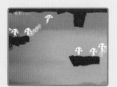

❾ WHERE'S MY WATER? (see pages 120–21)

Swampy the alligator really wants to take a shower. Unfortunately, his plumbing has more holes in it than Swiss cheese.

❿ DRAW SOMETHING (see pages 88–89)

Draw a picture of an item you've been given a description of. Sounds easy, right? Think again.

AMAZING ALEX

Category: Puzzle • **Available on**: Android/iOS • **Cost**: Free/Paid
Download size: Small • **Age rating**: Suitable for all

SYNOPSIS

Alex is bored, so what better way to dispel boredom than with incredible physics-based puzzles?

How do you follow a smash hit like Angry Birds? Well, if you're Rovio, the makers of Angry Birds, you do something completely different. Amazing Alex is a physics-based puzzle where you, as Alex, set up increasingly intricate contraptions. The game bears more than a passing resemblance to the board game Mouse Trap and also to an old PC game called The Incredible Machine. Think of it like a massive chain reaction or toppling dominoes. You begin with very basic setups but, as the game goes on, you'll find yourself using all sorts of household items in order to reach your goal.

GAMEPLAY

The game has a very clean, cartoony graphical style and everything is very easy to navigate. You'll be presented with the setup, which could be anything from getting a

football into a basket to trying to get pool balls through a series of pipes. The contraption isn't finished though, and you have an inventory full of the items you need to complete it.

This is done with a neat control method that makes putting your contraption together a doddle. You touch a button at the bottom of the screen to open the inventory and all the pieces available are presented to you. All that's needed is for you to drag the item from your inventory into the desired place on the screen. You can put it wherever you want and you're able to rotate it too.

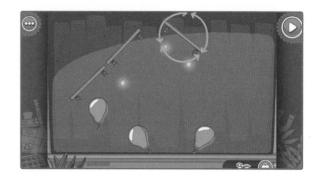

THE LOWDOWN

The levels can be completed by accomplishing the task set out at the start, but along the way you'll want to pick up stars to get the best score. Incidentally, the star-scoring system is the only thing in Amazing Alex that has been carried over from Angry Birds. Once you've got everything in place, and you think you've got it right, you press the play button in the top corner to see what happens. Very often you won't get it right, as that would be too easy and Amazing Alex is a tricky, challenging game. It has over 100 levels with more being added all the time, plus there's the facility to design your own levels and share them for others to play.

Amazing Alex is a worthy successor to Angry Birds. It's a completely different beast, but developers Rovio have shown once again their undoubted talent for mobile games.

You Might Also Like
Bad Piggies (pages 28–29), also from Rovio, takes the problem-solving of Amazing Alex and the carnage of Angry Birds and mixes it all together.

Hot Tip
Try every object at every possible angle. Don't be limited by what you assume won't work.

BEBBLED

Category: Puzzle • **Available on:** Android/iOS • **Cost:** Free/Paid
Download size: Small • **Age rating:** Suitable for all

SYNOPSIS

You have a screen full of coloured bebbles. They look pretty but, if you're going to win, they have to go. The best puzzle games are the ones where you don't need to read the tutorial or spend ages working out what to do. These are the ones that you instinctively know what to do as soon as you turn them on. Bebbled is one of those games.

GAMEPLAY

Bebbles are pebbles, coloured pebbles to be precise, and they cover the screen. There are different colours and you can get rid of bebbles by having two or more of the same colour together. Simply touch the screen and they vanish, leaving the remaining bebbles to shift across into their place. The shifting bebbles means that the game screen is always changing and new combinations appearing. The more bebbles you can make disappear, the higher your score. If you're running out of moves, then Bebbled has a surprise for you: simply turn your device upside down and the whole screen, bebbles and all, will flip and open up new combinations to you.

There are two modes available, the first of which is campaign mode, where the game sets you a certain number of points that you must score to reach the next level. Here, the key is to try to plan for combinations of five bebbles or more. Any fewer than that and you're likely to fall short.

The other is freestyle mode, in which there are six different game options, ranging from the classic continous mode, where you try to score as much as possible to move on to the next level, to the shifter modes. The shifter modes were our favourites; a combination of speed and relentless puzzling, they continue to add bebbles just as you think you're clearing the screen.

The best version here is TimeShifter, a nice twist on the beat-the-clock idea, where the clock stops as long as you keep playing but, if you pause to ponder your next move, the clock starts ticking.

THE LOWDOWN

Bebbled is a really cheerful game with crisp visuals and a lovely upbeat soundtrack, plus the bebbles have a nice colour palette. Both modes are good value and, with new campaigns being added, we'll be playing this one for a long time to come.

You Might Also Like
Drop 7 takes the horizontal/vertical puzzle format and gives it an interesting numerical twist.

Hot Tip
In the Christmas challenge mode, you only need to get rid of the glowing bebbles. Plan your strategy around that rather than removing every bebble on the screen.

CONTRE JOUR

Category: Puzzle • **Available on:** Android/iOS • **Cost:** Paid
Download size: Small • **Age rating:** Suitable for all

SYNOPSIS

Guide little Petit through five strange and abstract worlds by manipulating the beautiful, minimalistic landscape.

The greatest thing about mobile games is the resurrection of independent designers willing to take risks with their work. The end result is usually something so bold that it stands out from the crowd. Contre Jour (which means 'against daylight' in French) is that type of game. The name refers to the art style used throughout, with the background lit up but the foreground items, such as the land and your character, in silhouette.

GAMEPLAY

Your character, Petit, doesn't move directly. Instead, you move the ground around him. You raise or lower the land to move him along and direct him to the blue light that acts as the exit

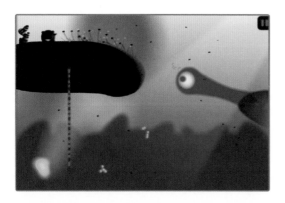

to each stage. Along the way, you collect floating blue lights which act as star ratings for your score, although you can complete levels without them.

As you progress, tentacles link to Petit to move him over greater distances. Black tentacles work like elastic and striped ones work like rope. You can attach more than one tentacle to give Petit more flexibility to navigate large obstacles or wide chasms. There are also slingshots, air pockets and portals that can be used to move Petit over taller obstacles. If you do get stuck, the option to skip a level is there. There are enemies on some levels that might trip you up and cause a restart, but they aren't that prevalent. The puzzles in Contre Jour are more about navigating the terrain than dodging the nastier natural elements.

THE LOWDOWN

The gameplay often feels like you're playing with Plasticine, as you roll surfaces back and forth to move Petit around each level. The graphics are a sight to behold and it often feels like you're looking at an inkblot drawing. The screen never gets difficult to navigate but, if you need to, the facility to pinch and zoom to get a closer look is there.

There's a wonderful piano score throughout Contre Jour, which complements its artistic sensibilities. The levels are structured in such a way that it's easy to dip in and out of, although we guarantee that you'll be drawn into Petit's strange and haunting world.

You Might Also Like
Cut The Rope (pages 86–87) has a similar style of play with a different graphical approach.

Hot Tip
You can hold tentacles out to catch Petit rather than trying to aim him at them.

CUT THE ROPE

Category: Puzzle • **Available on:** Android/iOS • **Cost:** Free/Paid
Download size: Small • **Age rating:** Suitable for all

SYNOPSIS

Feed candy to hungry monster Om Nom in this ingenious physics-based puzzle with dangerous levels of cuteness.

Om Nom is a green monster with a ferocious appetite, but don't let that put you off, because Om Nom is also one of the cutest little monsters you will ever see. In Cut the Rope you need to get Om Nom his favourite food, candy, by cutting the rope that it's suspended from inside Om Nom's box. It starts off very easy, as you're cutting one or maybe two ropes to deliver Om Nom his candy, but soon it becomes fiendishly difficult. You'll have to work out which way the momentum will take the rope and you begin to use items such as bubbles, automatic ropes, stretched ropes (which act like rubber bands), balloons and more. As the game progresses, you have to start cutting multiple ropes at the same time and avoid obstacles such as spikes, that destroy the candy, and spiders, who would very much like to eat it too.

GAMEPLAY

The gameplay is physics-based and it is very simple to use. Just cut the rope from which the candy is suspended and safely deliver it to Om Nom. Simply swipe your finger across the suspended rope to release the candy in the direction of the stars, which you want to improve your score, and of Om Nom, whom you need to feed to complete the level. As you progress, you unlock more levels, each with a different theme.

THE LOWDOWN

The music in each level is bouncy and suits the game's cartoony style. Om Nom is ridiculously cute and the graphics and animation are wonderfully simple but also very pretty, and his little face when you fail to feed him makes you strive to do better each time.

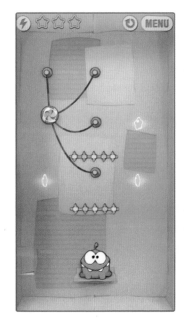

Addictive, hard to put down and suitable for adults and children alike, Cut the Rope is both free and insanely addictive. There is a paid version, which has no adverts, but in the free version they don't greatly detract from the overall enjoyment of making Om Nom a very happy little monster.

You Might Also Like
Cut the Rope: Experiments. More candy-chomping escapades with Om Nom as he's put through a series of tests to see just how much he loves candy. (Hint: it's a lot.)

Hot Tip
Remember that sometimes your reflexes have to be as sharp as your scissors!

DRAW SOMETHING

Category: Puzzle • **Available on:** Android/iOS • **Cost:** Free/Paid (with in-app purchases) • **Download size:** Small • **Age rating:** Suitable for all

SYNOPSIS

Draw a picture of an item you've been given a description of. Sounds easy, right? Think again.

The concept of Draw Something isn't just that you – for want of a better choice of words – draw something. It's about how badly wrong you can get it whilst trying your very best. You start by selecting an opponent. Like other games by Zynga, this can be done via Facebook, your contact list or looking up friends' usernames. Once you've set up a game, it's time to draw.

GAMEPLAY

You're given a selection of three objects, the difficulty of drawing each ranging from one to three stars. For instance, on one of our games, the one-star item was 'worm' but the three-star item was 'Bahamas'. Needless to say, we picked the one-star option on that one. If you're not happy with the selection of words you've been given, you can use bombs to give you a new one. The longer you and your partner can keep going, the better your streak will be.

Once you're done creating your masterpiece, you send it off for the other person to guess. The game offers a little assistance to the guesser, who's given the word size and a jumble of letters that belong to it. Be warned, not all of the letters belong to the word in question and you might have to use one of the in-game bombs to eradicate some of the erroneous letters.

As you play the game, you can earn badges by completing certain challenges, which are grouped together by theme. The more coins you earn, the more bombs and different colours you pick up as you go. If you want more colours even more quickly, you can buy different colours to spruce up your drawings. It's not necessary, and our artistic endeavours were never hampered by lack of colour, but it's a good option to have if you want it.

THE LOWDOWN

Draw Something can be played on almost any device, but we have to say the bigger the screen, the easier it is to exercise your artistic talents. It's not a big problem, but the more surface area you have to work with, the better.

Draw Something is one of the most fun social network games we've played. If you link it to your Facebook account, you'll be able to connect to your friends, which will ensure you'll always have someone to play with.

You Might Also Like
Any of Zynga's 'With Friends' games will fit the bill.

Hot Tip
Try to keep your streak going rather than passing.

FLOW FREE

Category: Puzzle • **Available on:** Android/iOS • **Cost:** Free
Download size: Small • **Age rating:** Suitable for all

SYNOPSIS

Join up your coloured pipes to keep the water flowing.

Flow Free is a brilliantly simple concept. Starting with the first level, you have a grid and a series of coloured dots. All you have to do is draw a line from one coloured dot to another on the grid. The lines you draw are meant to be pipes. They don't look like pipes but, quite frankly, that's really not important. There are water-themed sound effects, but you won't notice, as you'll be too busy linking your pipes together. The early levels are easy, but as you progress through the game things become much more challenging. Bigger grids start appearing with an even bigger selection of different-coloured dots. Soon you're playing on large grids with a huge assortment of pipes to connect.

GAMEPLAY

Control is nice and simple, and those of you with Android devices will realize that the game bears an uncanny resemblance to the lock pattern you use on your device. To draw a line from one coloured dot to another, you touch the dot and then drag your finger along the route you'd like it to take. If you make a mistake, just tap one of the coloured dots and the pipe you've drawn disappears so you can try again. Trust us, that will happen a lot in the later stages.

THE LOWDOWN

There are two ways to play, free play or time trial. Free play has a huge selection of levels (750 in total), so it's impossible to get bored. Another nice touch is the inclusion of a pack of levels optimized for tablets, and these are available at no extra cost. You can play the tablet levels on phones if you want, so no one is excluded. Time trial sets you the task of going through as many grids as possible within a time limit. You select the time limit from four options and you can even pick your grid size to maximize the challenge.

As an all-round package, there isn't much that comes close to Flow Free. It costs nothing and has a slew of options.

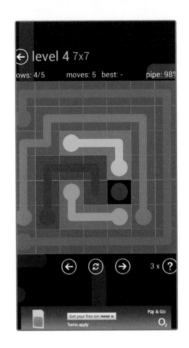

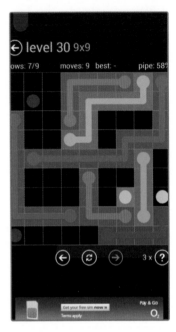

You Might Also Like
Flow Free: Bridges takes the idea of Flow Free and allows you to overlay pipes.

Hot Tip
Tackle the time trials before free play; it'll sharpen your technique before taking on the bigger grids.

MONOPOLY

Category: Puzzle • **Available on:** Android/iOS • **Cost:** Paid
Download size: Small • **Age rating:** Suitable for all

SYNOPSIS

The grand daddy of all board games flexes its muscles on your mobile device. Now, who wants to be the boot?

Everyone has played Monopoly at some point in their life, but now you actually get the chance to finish a game. The classic board game hasn't changed much in its move to mobile platforms, but now you can at least save the game if you don't have much time to play.

GAMEPLAY

If you want to play human opposition, you can do so in two ways. You can share a single device and pass it between you, or use separate devices connected via Wi-Fi or Bluetooth. Both methods are fun and neither detracts from the overall quality of the game. Controls are very clear and all your options are laid out at the bottom of the screen. When it's your turn to move,

you shake your device to roll the dice and away you go. You then have the choice of buying the property you've landed on or let it go to auction.

Buy properties until you have a complete colour group and then you can start making the big bucks by building houses and hotels and charging rent to other players. You can of course trade properties for cash and other properties; vital if you need to secure Bond Street before you can move the builders in. If things get really desperate, you can choose to mortgage some of your properties to gather in extra cash, but beware, as doing so will reduce your income when people land on your property.

THE LOWDOWN

The board is animated in 3D and looks beautiful, whilst the animation on the different pieces as they move along is quite charming. Everything is very clear and the colours are bright and vibrant. When you buy a property, a coloured tab appears below it so you'll always be able to keep track of your budding portfolio. The bigger the screen, the better the auction section works, but that's not to say it isn't very enjoyable however you decide to play it.

This Monopoly is a great spin on the classic board game and with this one, there's no chance of people tipping the board up when things aren't going their way.

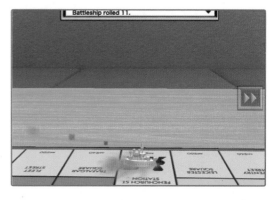

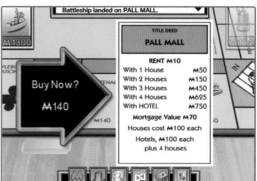

You Might Also Like
The Game of Life – do all the things you do in real life, but without the hassle.

Hot Tip
Try to undercut your opponents by taking properties to auction when they don't have a lot of money left.

THE MORON TEST

Category: Puzzle • **Available on:** Android/iOS • **Cost:** Paid
Download size: Small • **Age rating:** Suitable for all

SYNOPSIS

Think you're clever? Think again.

The Moron Test is a quiz with a difference. Obviously the title gives the impression that it's easy; it must be, because you're not a moron. Are you?

There are six different tests in this version, each with its own theme. There's Old School, which is the first version of the test, and then there are Late Registration, Winter Break, Food Fight, Skip Day and Tricky Treat. The themes are very loose and, although the early tests are modelled on school, none of them really sticks to that.

GAMEPLAY

The Moron Test is a simple one-touch game. It poses you a question and all you have to do is tap on the answer. The Moron Test isn't a quizmaster in the traditional sense though. Rather than ask you straight questions, the Moron Test revels in misdirection and wordplay. All of this is designed to make you feel like – you guessed it – a moron. Soon, you realize how the game is played and you become far less trusting of its methods. Suddenly you begin scrutinizing every question, looking for the trick component, and sometimes there is none. That's the Moron Test's greatest asset; it plays fast and loose with its own rules, which keeps you on your toes all the time.

If you get answers wrong, the game will send you back to predetermined checkpoints. These aren't infinite though, so if you continue to get things wrong, you'll run out of lifelines and soon you're back at the start. Sometimes questions aren't even questions. Sometimes they're just instructions, such as pressing a button or pressing objects in order. It might ask you to press an object, like a duck, in order of size, then later on it will ask you what you think is the same question again, until you realize that it has lulled you into a false sense of security and sneakily added in a little caveat that you hadn't noticed. The game conditions you to one thing and then plays on your complacency, which inevitably results in you getting it wrong.

THE LOWDOWN

This is a game for everyone, and it's really fun asking a friend to have a go and then watching them get flummoxed by seemingly easy tasks. It's also terrific value with six different tests that will keep you going for a while. The graphics are intentionally basic to fit the game's comedy-at-your-expense ethos. We guarantee that you won't get through it first time, but when you do, you'll be cursing your own stupidity for not doing it sooner, because you're not a moron. Are you?

You Might Also Like
The Moron Test 2 is also available if you're feeling particularly unflappable.

Hot Tip
Don't trust the game's own instructions.

PUDDING MONSTERS HD

Category: Puzzle • **Available on:** Android/iOS • **Cost:** Paid
Download size: Small • **Age rating:** Suitable for all

SYNOPSIS

The pudding monsters don't want to get eaten and the only way they can stop that is if they join together. Literally!

The pudding monsters are happy just chilling out in the fridge. The problem is they're also delicious, so the best solution is for all the different pudding monsters to join together and

Personalise your
Samsung Galaxy SIII

scare off the humans who want to eat them. This is your job, but there's one tiny flaw in the pudding monsters' plan. The pudding monsters are super slippery and, when they're trying to make their way through the world, the only way they can stop themselves moving is if they bump into something.

GAMEPLAY

You move your pudding monsters across the surface by swiping your finger in the direction you want them to go. If you swipe them across the screen and there's nothing there to stop them, like an ice cube or a flowerpot, they'll fly off the screen! The goal is to combine all the pudding monsters you have onscreen and, if you can finish on certain tiles, you'll score a star rating too.

As you go through the game, you'll encounter different types of pudding monsters. Some leave a trail that slows down the others when they slide across it, some move as a group and others are fast asleep and need to be woken up. All of them need to be combined to make a bigger pudding monster to complete the level.

THE LOWDOWN

The control system is very easy and the game is awfully addictive. You'll often find yourself about to put it down and then decide to have just one more go. The level design gets increasingly more intricate as the game progresses, but the game never puts in so steep a challenge that it would turn you away. You always feel like the solution is staring you in the face, even as pudding monsters fly off into the distance.

The sound and graphics are incredibly cute and it's no surprise this is made by the same people who brought us Cut the Rope (*see* pages 86–87). The action is very easy to get into and young kids will love the interface and quirky characters. For everyone else, this is another fun game that will keep you amused for ages.

You Might Also Like
Cut the Rope (pages 86–87) also takes cuteness to dangerous levels.

Hot Tip
You can show off your scores to a friend via Facebook and Twitter. You can then try to outscore each other.

THE ROOM

Category: Puzzle • **Available on:** Android/iOS • **Cost:** Free/Paid
Download size: Small • **Age rating:** 12+

SYNOPSIS

Turn off the lights, turn up the volume and prepare to dedicate your very existence to unravelling the engrossing mystery that is The Room.

Beware. Once you're inside The Room, you will not be leaving for a very long time. However, until you've picked apart this majestically woven, unfathomably intricate puzzle, you will have little desire to do anything else. As soon as you enter, the atmospheric, Tim Burton-esque soundtrack kicks in and you'll almost forget this is just a game. Rather, it becomes a very real mission. You're confronted with a safe, standing alone in a darkened room, illuminated only by a spotlight. It's your job to crack its many, varied secrets.

The story is thus. A brilliant scientist has entrusted you to follow up his work. Where is he? What has he done? What is so important only you can discover the truth? The answers are in the safe.

The further you progress through the four chapters, the more you'll learn and the more unsettling – dare we say scary? – everything becomes.

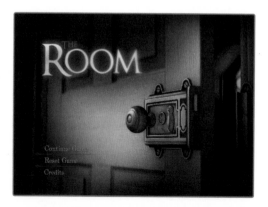

GAMEPLAY

The 3D world you are entombed within is marvellously interactive; you can pan 360 degrees, double tap to zoom in on objects and pinch to zoom out again as you search for clues. If you think you've looked closely for a clue, look again. You'll also need to employ gestures to turn keys, open drawers, line up patterns and rotate discs. It's all marvellously intuitive and an absolute joy to play. The countless puzzles you'll encounter are multi-layered and challenging, but still satisfyingly solvable. If you do hit a wall, the game will serve up handy clues.

THE LOWDOWN

The sequential nature of the puzzle is quite time-consuming, with each chapter taking around an hour to complete. For example, early in the game, you'll unlock a secret compartment to reveal a book; you must crack open the book to find a key. But where does this key fit and what's behind the keyhole once you find it? There are no quick answers.

Each discovery flows beautifully into the next and that's what makes leaving The Room nigh on impossible. It's by far the best, most beautiful, intuitive and rewarding puzzle game available for mobile devices, if not all gaming platforms. Are you up for the challenge?

You Might Also Like
Are you ready for the sequel? It's coming in autumn 2013.

Hot Tip
Your best friend is the eyeglass you discover very early on. Use it whenever you feel there is more to a puzzle than meets the eye.

SCRABBLE

Category: Puzzle • **Available on:** iOS • **Cost:** Paid
Download size: Small • **Age rating:** Suitable for all

SYNOPSIS

The classic word game comes to your mobile device, so there's no chance of losing tiles down the back of the sofa.

Scrabble is the rare instance of a much-loved board game actually being enhanced by its move to mobile gaming. No worrying about losing tiles or checking the dictionary for 'kwyjibo' here.

GAMEPLAY

There are three modes of play available. Quick play is for those who just want to get straight into it and show off their superior word skills against the computer. This mode also gives you Scrabble with all of the same rules you'd find if you played the original board game. Custom mode offers you the chance to tailor the game to your specific needs. Here, there are modes such as pass 'n' play, where players use the same device. This is a great feature for when there are two, three or four people, no Wi-Fi and only one device. You can also alter the difficulty level. If you're after a more target-driven game, then you can use the score-specific modes where the winner is the first to score a certain number of points, or you can limit the number of turns both players have in which to score. Both of these are excellent if you're pressed for time, such as playing on your commute to work.

Multiplayer Wi-Fi mode is the closest to classic Scrabble, allowing players to play each other over a Wi-Fi network. Whilst your opponent is deciding how to deploy all of those vowels they just picked up, you can shuffle through your own letters, planning your next move whilst keeping an eye on the board.

THE LOWDOWN

The display is bright, crisp and uncluttered. The game also helps you out by automatically zooming in on the board when you drag your tiles. This simple method means that you won't be for ever trying to get the best view of the board, although you can pinch and zoom if you wish.

Background music is unfussy, just as it should be when you're staring at your letter board and can't see where your next word is coming from. Luckily, the game doesn't leave you out on a limb; there's a 'Best Word' option that gives you the best word play, but beware, as there are only a few of these available, so use them wisely.

You Might Also Like
Words With Friends (pages 124–25) gives you the chance to take on the world with letters.

Hot Tip
Save those awkward letters for triple-letter scores, the underrated little brother of the triple-word score.

SCRAMBLE WITH FRIENDS

Category: Puzzle • **Available on:** Android/iOS • **Cost:** Free
Download size: Small • **Age rating:** Suitable for all

SYNOPSIS

Make words out of a grid of jumbled letters online against your friends and the rest of the world.

Zynga take another popular word game, in this case word search, and give it their social twist. You're given a grid of 16 jumbled letters and you have two minutes to make as many words out of them as you can. You play three rounds against another person and you earn coins as you go along.

Before you start your round, you can use different power-ups to play with. You use the coins that you've earned to buy these. The power-up options are Freeze and Mega Freeze (both stop the clock), Scramble (reorganizes your letters to give you more word options), Vision (gives you three words to find in your word jumble) and Inspiration (suggests words to play). If you're doing well, you can earn extra time to look for more words during each game.

GAMEPLAY

Control is by dragging your finger across the letters, and there's a satisfying clicking sound as you do so, making the whole thing feel incredibly intuitive. You make words by going forwards, backwards, horizontally, vertically and diagonally. In fact, anything goes apart from one little thing: all the letters that you select have to touch one another. With that rule, you can feel the game's origins in word-search puzzles, but Scramble With Friends is a lot more flexible as a game than that.

Each letter has points assigned to it, so you score by adding all the letters in your created words and then applying a multiplier, depending on the word's length. At the end of the round, the game will show you how many words you added and what your longest word was. There's also an option to show all the words that were available during your round from the letters you had. This is a nice little extra that gets you thinking bigger when it comes to the next round.

THE LOWDOWN

You link up via Facebook, email or by looking up friends' usernames. The username for Scramble With Friends also works with Zynga's other games, meaning you only have to give yourself one username that is used throughout the Zynga network.

Scramble With Friends is incredible fun for such a simple concept, which is what Zynga do best.

You Might Also Like

Both Hanging With Friends and Words With Friends are similar in style, yet different enough to be worth a look.

Hot Tip

Get 10 free coins when you sign up to Scramble With Friends through your Facebook account.

SHREDDER CHESS

Category: Puzzle • **Available on:** Android/iOS • **Cost:** Free/Paid
Download size: Small • **Age rating:** Suitable for all

SYNOPSIS

Pit your chess wits against one of the toughest chess programs ever. Only one can be the king.

Computer chess programs have been around for decades. The earliest PCs had chess games and, although the game itself is timeless, the computer opponent has become much fiercer.

Shredder

If you've never played chess before and can't tell your knights from your rooks, then Shredder Chess will help you feel your way through the different move set of each piece. If you're a casual chess aficionado (by that, we mean not terribly good at it), then the free version of Shredder Chess is for you. If you fancy yourself as a grandmaster, however, then the paid version will test you like never before.

GAMEPLAY

The game mimics a human player, meaning that it will work out your skill level and adjust itself accordingly. So you won't be getting beat downs constantly handed to you. Plus the game has a coach function that chimes in to tell you where you're going wrong and how you can improve. This isn't a game that wants to beat you. It wants you to learn and grow so that you become a better chess player.

Shredder Chess is completely no frills, there are no 3D boards or gimmicks here. It revels in its old-school sensibilities, going as far as using the picture of an old floppy disk for the save button. The control system is very easy to use; just select your chess piece and the game shows you all your available moves. The coach will also counsel you as you play.

There's also a challenge mode. Here, you're given a number of different scenarios to ponder and hopefully escape. Again, if you want a tougher task, there are more challenges available in the paid version, but we would recommend that you start with the free version and go from there. The game will also rate your ability against Shredder so you can monitor your progress.

THE LOWDOWN

This is a complete chess program for beginners and experienced players alike. It's so confident that it will give you a great chess experience that it doesn't mess about with unnecessary extras. Checkmate!

You Might Also Like
If you really fancy a challenge, take on Shredder in the full game. Not for the faint-hearted.

Hot Tip
Listen to your coach; he probably knows more than you do.

SONGPOP

Category: Trivia • **Available on:** Android/iOS • **Cost:** Free/Paid
Download size: Small • **Age rating:** Suitable for all

SYNOPSIS

Battle against friends to name that tune.

It's been a common debate in pubs and living rooms throughout the years. What song was that? About songs heard over the end credits of a film or on the jukebox in a noisy pub, that question has been asked a million times. Now along comes a game that challenges you to see just how good you are at guessing songs.

GAMEPLAY

Like all the best social games of this type, the idea is brilliantly simple. You hear a snippet of music and have 10 seconds to guess who the artist (or what the song title) is before the time runs out. You have a choice of four answers per song, so if you're not sure, you could always do what we did and guess. The game also offers you the chance to remove two of the answers that the game gives you, but use these carefully, as they're hard to come by once you've used them all.

The more quickly you answer, the better your score. If you guess consecutive answers correctly, then a score multiplier is added to boost your score. What's more, do well in a particular genre of music and points are added to your star rating for that genre. If you unlock enough stars, you'll get more tracks for that particular genre.

As always, for the impatient among you, there is the option to buy those extra genres.

THE LOWDOWN

Winning your battles against other players also means that, once you've earned enough, you can buy even more playlists, and there is a massive choice on offer. The selection screens are bright and cheerful and the sound quality on each song is particularly impressive.

You can link up with friends via Facebook, email, username or, if you just want a quickie, you can have a game against a random person. By the time we'd finished playing this, we had 12 different games going on. Sadly, it seemed our best genre was Classic Rock and our weakest was Today's Hits, officially confirming we'd turned into our parents.

You Might Also Like
Try your hand at movies instead of music with Who Wants To Be A Movie Millionaire? (pages 122–23).

Hot Tip
Although speed is important when answering, stringing together a row of correct answers is equally important. Take your time to listen to a bit more of the song if you're not sure.

SPIDER SOLITAIRE

Category: Puzzle • **Available on**: Android/iOS • **Cost**: Free
Download size: Small • **Age rating**: Suitable for all

SYNOPSIS

The card game that you've been playing all these years on your PC, now available to take with you.

Spider Solitaire has long been in the top 10 ways to procrastinate when you're at work. It's available on most people's work PCs, and who hasn't snuck in a game or two when no one's looking and you just feel like taking a break from work? There are plenty of Spider Solitaire games on mobile platforms, but we feel this version gives the best experience for your entire arachnid, card-related needs.

GAMEPLAY

When you start the game, you have a choice of card deals. The first is the Random deal. With this option, the game will shuffle the cards thoroughly and you aren't guaranteed to win. The other option is the Winning deal. Here there will be a winning hand in there somewhere, although you could still make a mistake and lose. The Winning deal option is useful because it also brings into play the 'Show me how to win' option, which is pretty self-explanatory. You can always have a previous hand dealt if you choose the Replay deal.

Moving cards from column to column is easy enough too. You can either select your desired card and drag it where you want, or you can double tap it and the game will move it for you. The scoring system is standard for Spider Solitaire and it's all displayed at the bottom of the screen, along with the time and how many moves you've made. The toolbar is down there too, but you can simply hide it by tapping the screen once and bring it back by doing the same.

THE LOWDOWN

The game screen is sparse but clearly set out. You can make the usual changes like different-coloured felt and different deck styles, but otherwise the game tries to make things as unfussy as possible. You can play it in portrait or landscape, and you alternate between the two by simply turning your device in the desired direction. It doesn't matter which way you hold your device, the game will automatically adjust itself to fit.

On the face of it, this game might look basic, but it has that essential quality possessed by solitaire – it's utterly addictive.

You Might Also Like
Try Zynga Poker (pages 128–29) for a different card-game experience.

Hot Tip
Make sure you give yourself a nickname so your high score can be recorded on the worldwide leader boards.

SPIRITS

Category: Puzzle • **Available on**: Android/iOS • **Cost**: Paid
Download size: Small • **Age rating**: Suitable for all

SYNOPSIS

Guide your little white spirits safely to the exit without letting any harm come to them.

Spirits are leaves come to life. They only want to get up into the sky and into the wind, where they can travel to the big white swirl and be blown around. If that all sounds a bit head in the clouds, don't worry. Spirits is basically along the lines of the old classic Lemmings done in the style of Japanese anime movies like those from Studio Ghibli.

GAMEPLAY

The aim of Spirits is to get them to the level's exit by various methods including digging, blowing them on a gust of wind, having them climb up on a vine or blocking a gust of wind that's in their way. You do this with the spirits themselves. These little white entities can be changed to do the tasks you require. For instance, if you need to get your spirits to higher ground, you turn one of your spirits into a fluffy cloud and he blows his friends up there. This is done by selecting your spirit on screen and the action you want from a pop-up menu.

Whilst the menu is open, your other spirits continue wandering about. This means you constantly have to think on your feet and be aware of your surroundings, because your spirits will go wandering into danger if you don't look after them. At the start, the game only requires you to save one spirit per level, but as you progress you'll have to save more. This means that you have to plan how you're going to manage your spirits.

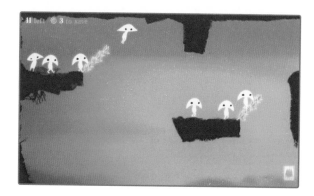

There are also flowers growing throughout levels and collecting these adds to your score. They're not always en route and collecting them and getting to the exit isn't straightforward, and it's worth taking a quick look around the level before making your next move.

THE LOWDOWN

This is one of the most fascinating and absorbing games we've played. The style of play, the art design and music all amount to a beautiful experience. There are lots of levels and plenty of challenges on each one, whether it be saving more spirits or collecting more flowers. Your spirits are such charming creatures that you become really invested in getting them to safety.

You Might Also Like
Contre Jour (pages 84–85) shares a similar design approach and is equally fun.

Hot Tip
Look out for the different wind currents on each level. These can guide you to hidden flowers for extra points.

SPRINKLE

Category: Puzzle • **Available on:** Android/iOS • **Cost:** Free/Paid
Download size: Small • **Age rating:** Suitable for all

SYNOPSIS

Sprinkle's home has always been peaceful, but space tourists have caused a bit of a problem and now only Sprinkle can save his people.

GAMEPLAY

Sprinkle is a simple concept but tough to master. Touch Sprinkle's fire crane to move it up and down, then touch anywhere else on the screen to adjust the angle at which the water will shoot out. Once you're all set with your angles, you press the red button on the bottom of the screen to release the water and get to work saving your people's houses. It's not just the angle or height of the water that's important, the layout of the level can assist you too. For instance, if you're aiming your water cannon upwards, you can let gravity help you, as any overflow rushes down to any fires below. If you want to score big, you'll need to be economical with your water. Using it all will get the job done, but your score won't be great.

As the game progresses, the levels introduce new environments and new obstacles. These can come in the form of boulders, blocks of ice and even cogs. All of them are vital to putting out the fires, be it as a bridge or as a weight to press a button to clear your path.

THE LOWDOWN

You can get a great experience out of the free version of Sprinkle, but if you want more of a challenge then there are more worlds and levels available in the paid version. There are plenty of levels to be getting on with, no matter which version you have, especially if you're trying to beat your own score.

The water in Sprinkle is not only well animated, as are the fires, but the gameplay is fantastic. The water always goes where you want it to go and the whole experience of putting out a fire is extremely satisfying. The music is jaunty and reminds us of the 16 Bit SNES/Mega Drive era, with a rhythmic musical accompaniment that reflects the tone and characters of the game. The animation is exceptionally smooth and the

blend of a 2D perspective and 3D objects works extremely well. There's also a nice little cartoon explaining the game's story, which is well worth a look.

You Might Also Like
In Where's My Water? (pages 120–21), Swampy the alligator needs a shower and you need to provide the water whilst avoiding the pitfalls of the sewers.

Hot Tip
Use gravity to save your water and get yourself a higher score.

+SUDOKU

Category: Puzzle • **Available on:** iOS • **Cost:** Free/Paid
Download size: Small • **Age rating:** Suitable for all

SYNOPSIS

Test your number-crunching skills without the need for an eraser or a newspaper.

Sudoku puzzles have had a meteoric rise over the last decade. They have moved from little-known time waster to the reason thousands of hours have been lost to their peculiar charms.

If you love Sudoku, this game presents you with something of a paradise: unlimited Sudoku puzzles. It's a surprise that there's no definitive game that straddles the needs of both Android and Apple users, but for us, +Sudoku is the best experience you can have on Apple devices.

GAMEPLAY

When you launch the game, you're presented with a blank grid. Select the difficulty you want and away you go. The game generates the numbers for each game, so the odds of you having a similar game to a previous one are remote. You can have up to nine games going at any one time, all saved and waiting for you to go back to. If you want to get rid of a grid or reset the timer, there's an option for that, and it makes changing grids a seamless experience.

The grid itself is optimized for touchscreens, so playing it is a dream. The layout sees your grid at the top and numbers one through nine at the bottom. Simply tap the square on the grid in which you want to input your number, and then tap the number below. It's that simple. The interface is easy to navigate and it's been designed with Apple devices in mind. It's so well integrated into Apple's own design scheme that it actually looks like it came with the device.

The game has customization options to adapt the grid to your individual tastes. You can have the game show up any conflicting numbers or highlight a number if you're looking for a certain combination.

THE LOWDOWN

The paid version has no ads (though we never found them intrusive) plus five extra difficulty levels of medium up to diabolical. We love Sudoku and +Sudoku is a brilliant way to pass the time. The ability to dip in and out means you'll always have something to do, no matter how much or how little time you have on your hands. Be warned, as once you start going, you just might lose track of time.

You Might Also Like
Drop 7 by Zynga takes the principle of Sudoku and fuses it with Tetris (pages 118–19).

Hot Tip
It's Sudoku, so we're sorry but you're on your own with this one.

SUPER MONSTERS ATE MY CONDO

Category: Puzzle • **Available on:** Android/iOS • **Cost:** Free/In-app purchases
Download size: Small • **Age rating:** 9+

SYNOPSIS

If you think the title is bizarre, wait until you get a load of the game.

Super Monsters Ate My Condo is mad, and we mean really, really mad. It's off-the-wall crazy and it doesn't care who knows it. All it wants you to do is have fun, oh and feed 100-feet-tall monsters, so nothing strange there.

A lot of love and effort has been put into making this game as weird and wacky as it can possibly be. The game borrows influences from sources like Godzilla monster movies, 1980s arcade games, Japanese Saturday morning cartoons and Jenga, and mixes it all together to produce a game that needs to be seen to be believed.

GAMEPLAY

The underlying premise is a simple one: you've got monsters and they want to be fed. It just so happens they've developed a taste for condos (flats or apartments to us). There are four monsters, each with outrageous names like Reginald Starfire and Boat Head, and each monster has a specific colour – red, blue, green or yellow. The condos are in matching colours and are stacked like a tower block. You can remove condos as you like with a swipe of your finger. There are only ever two monsters on screen at the same time and they like to be fed the condo that corresponds to their colour. Feed the wrong colour to the same monster too often and that monster will have a tantrum, toppling the tower and spelling game over.

There are power-ups to earn with each monster and you can switch monsters by stacking three condos of the same colour together. Each level you play has a set of requirements you have to meet before you can advance, and those requirements are as wacky as the rest of the game. You can earn coins as you go and they help unlock more power-ups such as score multipliers and special powers. There are also different types of condo that do different things to help or hinder your progress.

THE LOWDOWN

Beneath this bright, brash and intentionally bizarre game is a simple and addictive premise that is frantic and often very funny. The soundtrack is catchy and matches the action perfectly. It might look strange, but trust us, go with it and you will have an amazing time feeding the monsters.

You May Also Like
There's the original Monsters Ate My Condo or the even more bizarrely named Robot Unicorn Attack from the same developers.

Hot Tip
Use green monster's power to get two upgraded condos instead of one.

TETRIS

Category: Puzzle • **Available on:** Android/iOS • **Cost:** Paid
Download size: Small • **Age rating:** Suitable for all

SYNOPSIS

One of the most famous video games of all time returns to devour your time once again.

Tetris is rightly known as one of the most famous games ever made. Its most popular incarnation was on Nintendo's original Game Boy, where the game was packaged free with every console. This version is slightly altered from what the game was in its early 1990s heyday.

The original Game Boy was perfect for Tetris with its directional pad and two buttons. Now every device is touchscreen, so the developers have had to find a new way of playing Tetris.

GAMEPLAY

First, let's briefly explain how Tetris works. Different-shaped blocks fall from the top of the screen. You have to organize them as neatly as possible when they land. When you make a complete line, that line disappears and gets added to your score. The blocks fall increasingly quickly as the game goes on and the game ends when you have run out of space. You can also save blocks for later on in a game, but if you're an old hand at Tetris, you might consider that cheating.

There are three game options on this version of Tetris. There's One-Touch, Marathon and Galaxy. One-Touch is where the biggest move to adapt Tetris to touchscreen devices has been made. Here, the block appears at the top of the screen and there are shaded areas at the bottom where you can land it. Tapping the screen gives you different-shaded areas to choose from and, once you have, you tap the shaded area and the block moves there automatically. There's a timer on the side and if you exhaust this, the game dumps the block in the centre of the screen.

Marathon offers you the chance to start on whatever skill level you want with the old Tetris control method. Here, you drag and drop the blocks yourself, which still works well on touchscreens and is something for the Tetris purists.

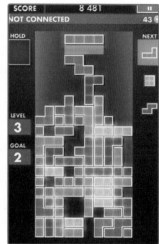

Galaxy is like a challenge mode, where the game asks you to clear certain blocks in as few moves as possible. It's a novel twist and it's a good break from the regular game.

THE LOWDOWN

Tetris is a classic game that will never go out of fashion, and the iconic theme tune and addictive gameplay are still very much present in this version. It's worth it for the nostalgia alone.

You Might Also Like
Bebbled (pages 82–83) is a game with its roots firmly in Tetris.

Hot Tip
If you're playing Marathon mode, saving up the vertical bars and building up your blocks will give your score a boost.

WHERE'S MY WATER?

Category: Puzzle • **Available on:** Android/iOS • **Cost:** Free/Paid
Download size: Small • **Age rating:** Suitable for all

SYNOPSIS

Swampy the alligator really wants to take a shower. Unfortunately, his plumbing has more holes in it than Swiss cheese.

There's no shortage of physics-based puzzle games available these days. But just because there's a lot of them, it doesn't mean they're any good. Thankfully, Where's My Water? is very, very good. The first thing that grabs you about this game is its presentation. It is spectacular from start to finish, which is what you would expect from a game that has been produced by Disney. The detail put into Swampy and the inhabitants of his world is mesmerizing, and it really feels like an interactive cartoon.

GAMEPLAY

Swampy wants to take a shower. Unfortunately, the plumbing to his comfortable sewer home has been severely damaged by another alligator called Cranky. We're not sure why this alligator-on-alligator abuse has been committed, but we blame Cranky's parents for giving him that name; he never stood a chance. Anyway, it's your job to get the water to Swampy using various and increasingly ingenious methods. Of course Swampy doesn't want to bathe alone, so along the way you'll need to pick up his rubber duckies to make his bath time more fun, and your score better.

At first, it's simply a case of clearing a path through the dirt to let the water flow, but soon it becomes clear that Cranky isn't going to make this easy. So you soon start coming across bombs, traps, dead ends, multiple routes and pipes, which, quite frankly, challenge all rational thinking about water in physics. The water doesn't last for ever, so make sure you don't let it all wash out into the wrong area, otherwise you'll have to start again.

THE LOWDOWN

All of this adds up to an incredibly enjoyable game. The controls are breathtakingly simple and incredibly intuitive. It's very simple to pick up and play, plus the colourful graphics and clever level design means that it will appeal to all ages; though we guarantee that kids will fall in love with Swampy.

There's plenty of challenge in the free or paid version, but the paid version does include over 200 levels if you're feeling ambitious. This game has been one of the most popular titles on mobile platforms and it's easy to see why. Cute characters, a great level of gameplay and lots of challenge add up to a must-have game. Just don't take it into the shower!

You Might Also Like
Sprinkle (pages 112–13) has the same emphasis on water-control physics.

Hot Tip
Look out for hidden items in the dirt on different levels.

WHO WANTS TO BE A MOVIE MILLIONAIRE?

Category: Trivia • **Available on:** iOS • **Cost:** Paid
Download size: Small • **Age rating:** Suitable for all

SYNOPSIS

The popular TV quiz goes Hollywood. Can you hold your nerve and win the top prize?

We love Who Wants To Be A Millionaire? (WWTBAM?) and, even though it isn't as popular as it once was, we still think it's one of the best quiz shows around. Different editions of popular quizzes like this are fine in principle, but they live or die by the strength and depth of their questions. We can remember the early mobile versions of Trivial Pursuit repeating the same questions after three games. There are no such problems here. Who Wants To Be A Movie Millionaire? has an astonishing array of questions, covering movies from the silent-film era all the way up to the present day. So if you think you're a real movie aficionado, you'll soon find out when you play this game.

GAMEPLAY

The presentation (music, graphics and lifelines) are all intact. All that is missing is the host. The game works exactly as the TV show does. You have 15 questions, four lifelines and two safety points where you win the minimum amount if you get the later questions wrong. The fourth lifeline is introduced after the sixth question and allows you to switch the question on screen. There's also an option to take the money and run. That's in there for authenticity, but you'll probably only use it if you're competing with a friend to see who can win the most.

Control is one-touch and the layout is very clean and crisp. The visuals are lifted straight from the TV version but are given a golden sheen to denote that this is the movie edition. You can post your score via Apple's Game Center against your friends, and scores are based on how far you get and how quickly you do it. It's a nice addition to have, but we don't think that will be the reason you'll keep on playing.

THE LOWDOWN

Who Wants To Be A Movie Millionaire? is a nice addition to the increasing stable of WWTBAM? editions. If you think you're the king of the pub quiz, then this is the game for you. It will test your knowledge and you'll have fun doing it.

You Might Also Like
The original version and two sport versions are also available if you want to test your knowledge in other fields.

Hot Tip
If you get stuck early on, Ask The Audience is the best lifeline to play; the computer audience rarely gets it wrong.

WORDS WITH FRIENDS

Category: Puzzle • **Available on:** Android/iOS • **Cost:** Free
Download size: Small • **Age rating:** Suitable for all

SYNOPSIS

Test your word skills in a familiar game with a new social twist.

Here's another Zynga game that not only encourages social interaction but is built on it. You can either log in with Facebook or, if you don't have an account, you can just log in with your email address. Logging in provides an easy way for you to start games with people you know and setting it up is no hassle.

Before you start playing, you can choose if you want to play against someone you know via Facebook, by searching for their username or by going through the contacts list on your mobile device. If none of that takes your fancy, then you can go for the pass-and-play option, with you and a friend sharing the same device.

GAMEPLAY

The game plays very simply. You're given an assorted number of letters and you have to make a word out of them. Each letter has a score assigned to it and the combination of your letter scores placed on special places on the board allows you to rack up points. Where Words With Friends has the edge over other word-based games is its social network capabilities. Not only can you play multiple games at once, but you can also chat whilst you're playing.

Graphically, everything has been kept simple, which is a wise move, as too much clutter on the screen would detract from a simple interface. Your letters are laid out at the bottom of the screen and to move them to the board you simply drag the required letter into place. Your score is also displayed at the bottom.

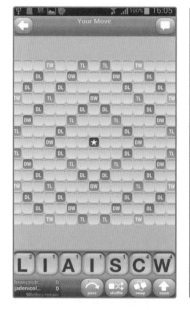

THE LOWDOWN

The game encourages you to have many games on the go at any one time, since your opponents might take a while to make their move. To start a new game at any time, whether against a friend or random, tap the plus sign at the top of the screen. The game always keeps you updated, so it's easy to keep track of your multiple games.

Words With Friends is a fantastic way to pass the time and its popularity means you'll always have someone to play against.

You Might Also Like
The obvious alternative to try would be Scrabble (pages 100–01), or maybe Hanging With Friends, a social take on hangman.

Hot Tip
Jumbling your letters gives a fresh perspective when you're stuck.

WORLD OF GOO

Category: Puzzle • **Available on:** Android/iOS • **Cost:** Paid
Download size: Small • **Age rating:** Suitable for all

SYNOPSIS

Video games and art combine in a witty puzzle that will absorb your time like a big ball of goo.

On the face of it, this is a fairly simple puzzle in which you need to move your inquisitive goo balls to the big pipe in the sky. But underneath this is a clever, inventive game that, when it's not testing your grey matter, will have you laughing at its deadpan humour.

GAMEPLAY

There are five worlds through which you must navigate your little balls of goo and there are multiple paths through each one. You'll want to go back and redo all the levels though, as there's so much to see and the detail in each level is worth investigating.

You have to build structures out of your little goo balls in order to complete a level. You score points according to the number of balls you were required to collect plus any extras, and by how many moves you did it in and how quickly. It's a simple formula, although the game is so pretty that you could argue that the enjoyment is in the playing rather than the scoring.

Any extra goo balls that you collect are sent to the World of Goo Corporation for use on a sandbox level (on which you can do and build what you want) to compete with others to see who can build the tallest tower out of goo balls.

THE LOWDOWN

The level of sophistication in the game's humour can be seen on all levels and in the mysterious Sign Painter's interjections. The Sign Painter is essentially the level guide, but it's performed with such a clever, knowing use of language that you never feel it's detracting from the experience.

The abstract graphics are intentionally minimalistic so that it almost seems more art-house film than popcorn blockbuster. Its use of music is also beautiful and at times it sounds like fairground music mixed with the work of Danny Elfman. There are also funny little animated segments in between levels that serve to enhance the game's plot.

World of Goo is an experience that will reward you every time you play it. It's an absolute steal for the price, as you'll probably never see anything so low in price but so high in quality. It's simply a wonderful, beguiling and beautiful piece of work.

You Might Also Like
Contre Jour (pages 84–85) is another game that treats its subject as art.

Hot Tip
Combine goo balls to get the job done faster on later levels.

ZYNGA POKER

Category: Puzzle • **Available on:** Android/iOS • **Cost:** Free
Download size: Small • **Age rating:** 12+

SYNOPSIS

Test your poker skills against others with this poker game that's sharper than a Vegas high roller.

Poker has been one of the most popular games of the last decade. Before the 2000s, poker was seen as something played by people in high-class casinos whilst wearing tuxedos. Now it's one of the most accessible card games out there. The rules are relatively simple, but poker is a game of wits, and if you're not careful, you could lose everything. Luckily, Zynga Poker isn't going to put you in any financial difficulty. The money on offer here isn't real, so the only thing you'll lose, if you go bust, is your pride.

GAMEPLAY

Zynga have taken their tried and trusted method and applied it directly to poker. That is to say that they have accentuated the very best social aspects of it and minimized the fuss in setting

up a game to play. We're not going to go into all the different types of poker hands here, but if you're unsure, they're in the help section under 'Hand Ranking'.

As usual with Zynga games, you can log in with Facebook or with a Zynga account that you have no doubt already created. If not, just put your email address in and your account is set up for all Zynga games. It's possibly the best aspect of Zynga's social gaming platforms. The setup around the poker table itself is straightforward. You're always placed at the bottom to avoid confusion. There's also the option to invite friends to your table if there's a spare seat. There's a chat option too, if you want to rub your opponents' faces in it when you're winning.

THE LOWDOWN

The game itself is well presented and all the sounds of cards being dealt and chips falling are there to recreate that authentic casino atmosphere. Your options to fold, check, call and raise are all laid out below the table. If you choose to raise, then a sliding bar appears on the side for you to pick your amount.

We lost countless hours on Zynga Poker and it's a great way to pass the time. It's simply the best social poker game out there.

You Might Also Like
If you're in the casino mood, why not try Zynga Slots and try to win big?

Hot Tip
Sometimes it's better to play a hand even if it's not a winner. This throws people off your scent when you're bluffing later on.

SPORTS & RACING GAMES

SPORTS & RACING GAMES

Sports and racing games continue to be among the most enduring gaming genres of all and, let's face it, it's not difficult to see why. We'd all love to score the winner in the World Cup final, or blaze our way to victory in the Monaco Grand Prix. Thankfully, mobile game developers can help us virtually fulfil those dreams.

EA Sports has long led the way for faithful, hyper-realistic sports games and that tradition continues with simulators like FIFA Soccer, Tiger Woods PGA Tour and Madden NFL for iOS and Android devices. SEGA's Virtua Tennis Challenge is also a brilliant touchscreen-friendly game you simply must try.

FUN AND GAMES

However, not all sports games within these pages are focused on grandstanding visuals and realistic gameplay. Where's the fun in that? Cartoonish slog-a-thon Stick Cricket might be the most pleasing of the lot, while single-finger, arcade-style game Flick Golf is hugely enjoyable for nonperfectionists. Also far detached from reality is the beloved two-on-two basketball game NBA Jam, which gets a new lease of life on the touchscreen. And, for those nights when you can't get down to the local, Darts Night and Pool Break are ample substitutes.

Above: Not all sports games have hyper-realistic visuals. NBA Jam is far detached from reality.

There are also some fine strategy games within these pages. Football Manager Handheld 2013 allows you to take the hot seat and guide your club to untold glories, while career-based New Star Soccer and Baseball Superstars require you to build your skills up over time by performing well on and off the field.

AS GOOD AS IT GETS

As far as racing is concerned, the highlights are console-quality, visually stunning games like Asphalt 7: Heat, Real Racing 3 and Need For Speed: Most Wanted. With their brilliant tilt-to-turn controls, endless tracks to master and cars to unlock and race, they're the pinnacle of the genre.

However, we've also tried to mix up the modes of transport a bit. You can get behind the wheel of a speedboat in Riptide GP, guide a steam-powered death machine in Steampunk Racing 3D, rely on grappling hooks to catapult your vehicle in Slingshot Racing, and take to the slopes in Crazy Snowboard.

Above: Arcade-style Flick Golf is great fun for golf enthusiasts.

HANG ON FOR THE RIDE!

Strangely, the object of some of these racing games isn't always just the winning. In Moto X Mayhem, you'll need to finish the course in the fastest time possible, while somehow staying on the bike, while Highway Rider requires you to earn points by dodging in and out of traffic. In the endlessly fun Reckless Getaway, you have to avoid capture at the hands of homicidal police forces that are trying to run you off the road. Stay out of trouble, you crazy kids.

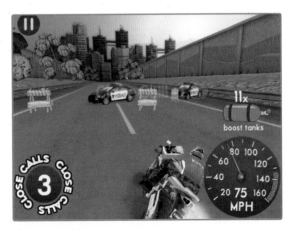

Above: Highway Rider is a racing game with the added challenge of having to weave in and out of traffic.

CHRIS'S PICKS

We hope you enjoy playing these games as much as we did. For the record, here's my personal Sports and Racing Top 10:

❶ ASPHALT 7: HEAT

(*see* pages 136–37)
A pedal-to-the-metal arcade-racing classic. Because who needs breaks?

❷ FIFA SOCCER 13 (*see* pages 144–45)

The Lionel Messi of football simulators comes of age on mobile devices and the beautiful game has never looked better.

❸ NBA JAM (*see* pages 158–59)

He's heating up ... HE'S ON FIRE! Relive the best basketball game ever.

❹ STICK CRICKET (*see* pages 176–77)

A simplistic but endlessly fun cricket game that hits the competition for six.

⑤ FLICK GOLF (see pages 146–47)

Leave your clubs at home and tee off with your flicking finger, earning points and unlocking new courses around the globe.

⑥ NEED FOR SPEED: MOST WANTED (see pages 160–61)

Racing against a host of street-savvy speedsters is tough enough, but with cops looking to run you off the road, things can get tricky.

⑦ FOOTBALL MANAGER HANDHELD 2013 (see pages 148–49)

Real-life managers say it's a 24/7 job. Don't expect this game to be any different.

⑧ VIRTUA TENNIS CHALLENGE

(see pages 180–81)

Great controls, fabulous graphics, plenty of tournaments and fast-paced gameplay. What's not to love?

⑨ SLINGSHOT RACING (see pages 172–73)

Slingshot Racing reinvents the wheel with a near-perfect, supremely original racing game.

⑩ CRAZY SNOWBOARD

(see pages 140–41)

Almost as much fun as real snowboarding, but without the broken bones.

ASPHALT 7: HEAT

Category: Racing • **Available on:** iOS/Android • **Cost:** Paid
Download size: Large • **Age rating:** Suitable for all

SYNOPSIS

A pedal-to-the-metal arcade-racing classic. Because who needs breaks?

Channelling the spirit of the sit-down arcade machines of yesteryear, Asphalt 7 places you behind the wheel of one of up to 60 fully licensed and beautifully realized sports cars and lets you slam them around stunning real-life city locations, such as London, Miami and Rio de Janeiro. Your aim is to win races and trophies and unlock all of the cars and tracks the game has to offer.

GAMEPLAY

Gameplay is quite simple to master, making it easy for beginners to buckle up. You don't control the accelerator pedal; you just hit the breaks by tapping the left of the screen, while tilting left and right controls the steering (although that'll soon become a full-body movement!).

The action is fast-paced enough, but you can gain extra speed by building up your turbo power, which you do by grabbing icons littered around the road. If your turbo is fully charged when deployed, the screen turns blue and you become utterly unstoppable for a few seconds. As we said, this is an arcade-style game rather than a simulation, so if you wreck yourself, it'll cost you a few seconds, but there'll be no damage to the car.

There are varying challenges to compete in: regular races where you have to finish ahead of the pack, elimination races, time trials and others like the awesome knockout challenges where you must wreck the opposition's cars. As you complete races and goals, you'll earn stars and money. The stars can be used to unlock new cars, while the money buys them. You can cheat and unlock cars with real money, but there's plenty of mileage in playing the career mode.

THE LOWDOWN

Lastly, can we please talk about the looks? If you're lucky enough to have an iPhone or iPad with the Retina Display, then there are few games that'll look better than Asphalt 7.

The visuals and flawless, high-speed rendering of the tracks are beyond compare and it wouldn't look out of place on an Xbox 360 or a PlayStation 3 console. To add to the fun, you can play with friends via local Wi-Fi and Bluetooth and even sign up for an account to take on the online community.

You Might Also Like
Check out the Asphalt back catalogue, and also EA's Need For Speed series. Reckless Racing 2 and Real Racing 3 are also crackers.

Hot Tip
For a really immersive experience, hit the camera icon during a race to get a driver's-eye view.

BASEBALL SUPERSTARS 2013

Category: Sports • **Available on**: Android/iOS • **Cost**: Free (with in-app purchases) • **Download size**: Small • **Age rating**: Suitable for all

SYNOPSIS

Part sports, part role-playing game, this anime-inspired baseball game is a bona fide home run.

Professional athletes are always trotting out the old cliché: 'Success is as much about what you do off the field as what you do on it.' However, in the case of the highly addictive Baseball Superstars 2013, this adage holds true.

You're challenged with working your way up to superstar status by performing well in games, hitting home runs or throwing strike outs, but to do that you'll need to train hard to improve your stats, spend wages on better equipment, and find the right balance between practice and relaxation. You can choose between becoming a pitcher or a hitter, but hitting is the most fun, so we'll focus on that. Pick your team and set yourself goals for the season (home runs, hits, on-base percentage) and you're ready to go.

GAMEPLAY

The cartoonish gameplay is simple. When the pitcher throws, you'll need to time your swing perfectly, otherwise you'll be struck out. Make a bad contact and you'll be run out or caught in the field. It's one touch of a button but it's harder than it sounds, because pitchers have plenty of variation, while the 'super pitchers' are nigh on impossible to hit. If you play well, you'll earn 'active' points, which can be spent in the yard on improving your skills or resting to improve your conditioning. But you can also visit the park, where you can take on a part-time job, check into hospital to treat injuries and visit the shop to buy new equipment. You'll also need to talk to people to make friends, including the weird ghost-girl who hangs out in the yard. Indeed, this game is as much role-playing as it is sports.

THE LOWDOWN

At the end of the season, your performance is evaluated. You'll set new goals, hopefully as a better player, guiding your team to success. You'll need to put in a few seasons to get the best from this game, but it's worth it. We loved Baseball Superstars, and its anime-inspired graphics and characters make it a lot more fun than some of the more life-like options.

You Might Also Like
Big Win Baseball allows you to curate your very own dream team, while 9 Innings: 2013 Pro Baseball amps up the realism. Both are free.

Hot Tip
As strange at it may sound, if you're in a rut, turn the sound off and try again. It helps focus your mind on the ball.

CRAZY SNOWBOARD

Category: Sports • **Available on:** iOS/Android • **Cost:** Free/Paid
Download size: Small • **Age rating:** Suitable for all

SYNOPSIS

Almost as much fun as real snowboarding, but without the broken bones.

There are quite a few extreme sports games on iOS and Android, but if you'd prefer not to wipe out every five seconds, then Crazy Snowboard is for you. The gameplay is very straightforward. You control your boarder's forward motion by tilting your handset to the left or the right. Stick to the racing line and you'll travel faster. Likewise, if you go wide, your progress will be slowed. Hit an obstacle, or fail to land a jump, and you'll crash. But don't worry, snowboarders are tough and he gets right back on again.

GAMEPLAY

The real fun comes from executing various jumps and tricks. Once your rider gets some air, he can perform a host of flips and board grabs using the on-screen buttons. Hold down the button until the trick is complete and then your rider straightens out. The more air you get from ramps littered around the slopes, the more time you have to pull off tricks.

The idea is to guide your rider through various missions, grinding rails, navigating through slalom gates, collecting coins, all before the time runs out. However, the really fun missions require you to amass a certain number of points in the allotted time. This means you'll need to attempt more tricks and build up multiplier points as the seconds tick away. It's great fun.

THE LOWDOWN

The free version of the game has some simple missions, which you'll knock off pretty much straight away, but the full version (only around the price of a chocolate bar) gives you 30 missions to complete. You can also unlock new characters and better boards. Once you've completed all of the missions, there's the incentive to complete them again, and rack up all three stars.

It's one of the most graphically stunning games you'll see on a mobile device. You can be forgiven for losing track of your tricks while you admire the mountain scenery. All in all, Crazy Snowboard is a fabulously well-rounded game, with enough to challenge more advanced gamers and plenty for beginners to enjoy.

You Might Also Like
The legendary Tony Hawk's Pro Skater 2 is available for iOS only, while the iStunt games, where you tilt the phone to perform tricks, are also pretty fun.

Hot Tip
Hold your grabs for as long as you can to boost your scores, and make sure you grab the purple crates – they're worth extra points.

DARTS NIGHT

Category: Sports • **Available on:** iOS • **Cost:** Free/Paid
Download size: Small • **Age rating:** Suitable for all

SYNOPSIS

It's scoring for show and doubles for dough. Ladies and gentlemen, LET'S ... PLAY ... DARTS!

If there's a sport that requires greater precision than the noble art of darts, then we're yet to discover it. However, we tend to struggle when attempting to match the greatness of Phil 'The Power' Taylor when chucking from the oche. So it is that we've found more joy pounding the treble 20 on an iPhone screen with a simple flick of the finger.

GAMEPLAY

Darts Night is a first-person game, meaning you get the player's-eye view as you take your aim. The idea is to flick upwards from the bottom of the screen in order to release each of your three darts. You'll need to get the direction and the weight of the throw (flick) right in order to hit your target. In the traditional game mode, you chip away at your score with each turn (starting from 301 or 501 and working down to zero), and win by hitting the double that corresponds with the score you have left. For example, if you have 20 points left, you need to hit the outer rim of the 10 to claim victory.

The 'flicking' technique takes a while to get used to and you may find yourself overshooting the target and missing the board a few times, but you'll soon get the knack and start hammering in those 180 maximum scores and taking out the big finishes. The trouble is, you have an opponent who isn't going to wait around for you to up your game.

THE LOWDOWN

You can play against the computer in a series of tournaments and challenges (like

Round the Clock) that become more difficult as you progress through the ranks. You can also play against friends over Wi-Fi or by passing the device back and forth. If you're struggling to find an opponent at your 'local', you can find a willing challenger online. It's ad funded, so the game is free, but you can get rid of the commercials with an inexpensive in-app purchase.

You Might Also Like

Sadly, Darts Night is an iOS exclusive at present, but Android gamers can get in on the arrows action with titles like the highly rated Darts 3D.

Hot Tip

Try to release the dart just as your finger reaches your target and don't rush your throw. Holding the phone straight also helps with accuracy.

FIFA SOCCER 13
FIFA SOCCER 12 FOR ANDROID

Category: Sports • **Available on:** Android/iOS • **Cost:** Paid
Download size: Large • **Age rating:** Suitable for all

SYNOPSIS

The Lionel Messi of football simulators comes of age on mobile devices and the beautiful game has never looked better.

FIFA Soccer is, quite simply, one of the most-loved video games in history. Indeed, we've been playing it since we were swapping Panini stickers in the school playground. Little did we know in those halcyon SEGA Megadrive days that almost 20 years later we'd be playing on a touchscreen mobile telephone and enjoying it just as much! As the years have moved on, so has FIFA Soccer and, after some underwhelming performances on mobile, the latest versions (12 and 13) are finally living up to the hype.

This fully licensed title still brings all of the real players, real-life tournaments and stadiums, intelligent in-game commentary and authentic crowd songs, but now the graphics, gameplay and overall fluidity of the experience have also reached the standard we'd expect from a FIFA Soccer game, and it's been tailored well to the mobile experience.

GAMEPLAY

In place of a physical controller, the game features a virtual directional pad, allowing you to move players with and without the ball, while the acts of passing, shooting, tackling and sprinting are facilitated by the on-screen buttons. There's also a host of skill moves you can use to dazzle your opponents. It's surprisingly easy to get the hang of it.

Beyond the Quick Match and Career Mode's countless leagues and tournaments, you can also enter the Manager Mode, which sees you take care of the club's finances, contracts and transfers and deal with board expectations. A new addition in FIFA Soccer 13 is the ability to play online against friends and random opponents, and you can also upload replays of your favourite goals to YouTube for bragging purposes.

THE LOWDOWN

EA Sports has chosen not to release FIFA Soccer 13 for Android gamers. It's a shame, but FIFA Soccer 12 is still an excellent alternative and miles ahead of the chasing pack.

At 1.48 GB, it will take up quite a lot of space on your phone/tablet's memory and takes ages to download, but it's definitely worth it. It's the best football game ever on mobile devices and you'll be playing until FIFA Soccer 14 comes out.

You Might Also Like

If you're looking for a less realistic approach to the beautiful game, try the likes of Flick Soccer, while the career-centric New Star Soccer (pages 162–63) is definitely worth a try.

Hot Tip

When you're going for goal, swipe up as you press the shoot button to deftly chip the goalkeeper.

FLICK GOLF

Category: Sports • **Available on**: iOS/Android • **Cost**: Paid
Download size: Small • **Age rating**: Suitable for all

SYNOPSIS

Leave your clubs at home and tee off with your flicking finger, earning points and unlocking new courses around the globe.

If Tiger Woods PGA Tour 12 (*see* pages 178–79) is St Andrews, then Flick Golf is your local pitch-and-putt, but that doesn't make it any less fun. Gone are the complex club choices, undulating greens and pristinely animated characters. Here, all you need is your trusty flicking finger and a good eye for a hole in one!

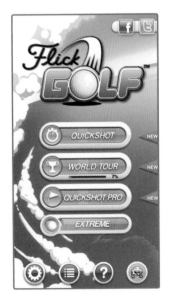

GAMEPLAY

You get one shot to get as close to the hole as possible – preferably in it – by flicking through the ball from the bottom of the screen. The power of the flick determines how far the ball travels, while aiming to the left or right of the flag counteracts the direction and strength of the wind. Once you've sent the ball flying towards the pin, you can swipe in any direction to control the spin. If you think you've overshot, start flicking down to apply backspin. Too far left? Add a little sidespin to hook it towards the hole. You can continue to control the spin once the ball has bounced in order to get it closer to the pin, and therein lies the secret to holes in one.

A target surrounds the hole, and the nearer you come to the pin, the more points you earn. There are spin bonuses and extra points for hitting the pin. If you score an 'ace', your points go way up. Hit multiple, consecutive holes in one, and you're in golfing dreamland. Land in the drink, however, and you'll lose 1,000 points. As you progress, you'll unlock new venues where the wind is stronger and there are more hazards to overcome. You'll also need higher points tallies to complete the course. By far, the most fun gameplay mode is Quickshot, where you're competing against the clock. Hit the red zone around the pin or hit a hole in one and you'll add more time to the clock, giving you more shots.

THE LOWDOWN

Gameplay is super simple and highly addictive. Graphically it's beautiful, especially on high-resolution screens like the iPhone 5. There are plenty of courses to experience and you can pay a small in-app fee to unlock them all, but it's definitely more fun to do it by achieving the targets.

You Might Also Like
Mastered all courses? Then get Flick Golf Extreme and tee off from icebergs and fighter jets.

Hot Tip
While it's fun to land the ball next to the pin, it's also very difficult. You'll enjoy more success by slightly overshooting and using backspin.

FOOTBALL MANAGER HANDHELD 2013

Category: Sports • **Available on:** iOS/Android • **Cost:** Paid
Download size: Medium • **Age rating:** Suitable for all

SYNOPSIS

Real-life managers say it's a 24/7 job. Don't expect this game to be any different.

Seasoned veterans will tell you, it's been a dangerous business, starting a game of Football Manager. People have lost entire summers as they ponder training routines, formation tweaks and new signings to assist them on the quest for domestic and European domination. Thankfully, the game is now available for mobile devices for those times when a 24-hour-a-day habit isn't conducive to the demands of a normal person's work/social life.

GAMEPLAY

For the uninitiated, Football Manager Handheld 2013 sees you take control of every aspect of managing a professional football team (choose from hundreds of fully licensed teams in 14 countries). You pick the team, carefully select tactics, run training sessions, deal with the media,

negotiate staff contracts and buy and sell players. However, true to life, once the game is underway, there's only so much you can do to control the outcome, which is agonizingly played out by the computer, with text commentary and a simple visual representation of the action.

If you're winning, you can switch the formation or bring on an extra defender. If you're behind, you can go gung-ho in search of a late equalizer. You'll savour your great triumphs and mourn those heart-wrenching losses for days. A run of good form increases your standing among players, fans and directors. You may even be offered your dream job at a Premier League club. However, fail to live up to expectations and, just like in real life, those trigger-happy chairmen will be handing you your P45.

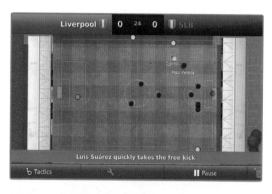

THE LOWDOWN

Gameplay is smooth and loading times are fast (especially compared with the first PC versions of the game!) and while the level of detail isn't quite as exhaustive as the desktop/laptop version, that's probably for the best. We've lost count of how many times we've missed our bus stop while masterminding triumphs on the way home from work.

The mobile version also brings a handy short-term Challenge mode, where you can, for example, take over a struggling team with the goal of saving them from relegation, making the game more suitable for those who would rather dip in and out.

You Might Also Like
Online Football Manager is a decent free alternative, and you can play against friends by signing in with your Facebook account.

Hot Tip
Keep an eye on your players' match fitness and morale. These two factors can often have the biggest effect on their performance levels.

HIGHWAY RIDER

Category: Racing • **Available on:** Android/iOS • **Cost:** Free (with in-app purchases) • **Download size:** Small • **Age rating:** 12+

SYNOPSIS

Racing games aren't often associated with comedy, however this simple yet brilliant daredevil title had us laughing out loud.

Promise us, motorcycle enthusiasts, whatever you do, don't play this game before strapping on your leathers and hitting the open road. We'd hate for the unquenchable thirst for danger that Highway Rider unleashes to spill over into the real world. This arcade-style racer deals in the currency of Close Calls as you speed towards your inevitable demise. You're encouraged to squeeze through the smallest of gaps, virtually scraping coats of paint off fellow road users, by tilting your device. The more daring the Close Call, the higher the reward.

GAMEPLAY

A regular Close Call earns one Gas Cap, a Show-Off move earns two, a Mad-Man manoeuvre earns four and a Perfect earns five. These are racked up throughout the game and can be spent on unlocking new characters (which you can also buy outright with real money). In the two game modes, Arcade and Fugitive (where you need to avoid police road blocks), the action gets faster and faster, while tapping the screen spends turbo boosts, leading to some seriously frenetic gameplay. 'Where are the brakes on this motorcycle?' we hear you ask. Well, there aren't any.

Naturally, the exponentially increasing speeds and the intentional courting of danger leads to incredible crashes. You'll fly up in the air, performing more somersaults than an Olympic diving champion and landing in a crumpled mess like a leather-clad ragdoll. The game will blithely list your injuries and your estimated medical expenses, but thankfully these don't affect your ability to climb back on the bike for another try. If you're bored of breaking your own records as well as your bones, you can play a friend or a random opponent online, where you're challenged with outlasting them and earning more Close Calls. It's fun and adds a competitive edge to the game.

THE LOWDOWN

Graphically, Highway Rider is decent enough, but it's the addictive gameplay that had us coming back over and over again. Many racing games require a huge time investment, but you can enjoy Highway Rider in bite-size, carnage-filled intervals.

You Might Also Like
Bike Baron is a great stunt motorcycle game that allows you to create levels of your own, but sadly it's iOS only.

Hot Tip
If you see a car in front indicating, head in the same direction. The chances of death increase dramatically, but you'll earn more points for your Close Call.

MADDEN NFL 12

Category: Sports • **Available on**: Android/iOS • **Cost**: Paid
Download size: Large • **Age rating**: Suitable for all·

SYNOPSIS

EA Sports' Madden games have been around for 25 years. Now you can carry the action around in your pocket.

American football isn't for everyone, but for gridiron fans and those who just enjoy a little competitive sporting action, there are plenty of options. For the full NFL experience, the officially licensed Madden NFL 12 is still the best.

GAMEPLAY

The idea of the game is to score as many touchdowns as possible on offence, and to stop the opposition when on defence. Naturally, trying to score is more fun. When your quarterback gets the ball, you have to decide what to do. You can hand it off to a running back, who can

beat the opposition with power and speed, or toss it downfield to a wide receiver. You make the choice by tapping the players. When they receive the ball, you can use the D-pad to control their direction, while hitting the 'sprint' and 'juke' buttons helps you evade tackles as you make for the end zone. If you don't score, the ball is handed over to the opposition and you have to stop them by rushing the quarterback, tackling the ball carrier, or intercepting passes. It's a lot more of a challenge on defence, because more often than not your teammates do the work for you! Once you figure out the control system and acquaint yourself with the weird rules, there'll be no stopping you.

To avoid playing on defence completely (which can be quite unfulfilling), select the fun TD Challenge mini game, which gives you two minutes to score in order to win the game. It's great when you have a small window in which to play. However, if you're in for the long haul, you can play a full season and lead your team to the Super Bowl championship. Multiplayer action is also available through EA's Origin platform, but if you have a friend nearby, you can play each other on two devices.

THE LOWDOWN

With so many players on the field and quite precise touchscreen controls, Madden NFL 12 is a much more pleasurable experience on tablets, but that's not to say the small-screen version is a bust. This is a realistic, great-looking version of the console classic.

You Might Also Like
NFL Pro 2013 is free to play, but lacks the gloss and playability of Madden NFL 12, while Backbreaker Football and NFL Rivals are fun and simple arcade-style games.

Hot Tip
Generally, mobile gamers don't have the time or inclination to spend hours poring over the playbook, so switch on Game Flow to auto-select plays.

MINI MOTOR RACING

Category: Racing • **Available on:** Android/iOS • **Cost:** Free/Paid
Download size: Medium • **Age rating:** Suitable for all

SYNOPSIS

Perhaps the most charming, playable and enjoyable little racer you can play on your mobile device.

There are loads of racing games that look better, boast more refined gameplay and offer faster action than Mini Motor Racing, but very few are as much fun. In this top-down racer (meaning you view the action from above, rather than behind the car), you'll take the wheel of a dinky little vehicle and must recklessly rip it around a host of tight little circuits, smashing into walls, careering into opponents and turbo-boosting your way to victory.

GAMEPLAY

You control your vehicle using a unique on-screen wheel that you hold down and spin, in order to make your turns. It takes some getting used to and it makes for an intentionally ragged

racing experience that contributes to the game's endless playability. It also makes a change from some of the tilt-to-turn racers already featured in this book. (You can change the steering method to tap or slide, but it's not as much fun.) Speaking of endless playability, the career mode is a little more conventional. Winning and performing well in races will earn you virtual cash, which can be used to improve your car or even to upgrade to new, faster models. Each car has its own driving style, so try a few and pick one that suits you.

The secret to success is to learn the tracks, master the racing line and know the best times to light up those tyres with the nitro boosts you'll find littered across the track. As you progress, you'll unlock more than 100 races on 20 tracks and race on a number of terrains in various challenging weather conditions. There really are hours of fun available here in a dip-in-dip-out capacity.

THE LOWDOWN

There's a free version of the game for you to try before you take the plunge with the full version (which won't greatly trouble the bank balance). The full version offers a neat multiplayer mode that allows players to beat down their real-life friends or random individuals online.

You Might Also Like
Reckless Racing, from the same stable as Reckless Getaway (pages 168–69) is an obvious and highly entertaining alternative.

Hot Tip
This is one of the few games that places you at the front of the starting grid. Take advantage and use a nitro boost to shoot off the line and gain an early lead.

MOTOX MAYHEM

Category: Racing • **Available on:** iOS/Android • **Cost:** Free/Paid
Download size: Small • **Age rating:** Suitable for all

SYNOPSIS

This endearing skill game is easy to play, but tough to master.

OK, we'll admit it. We were very close to giving up on MotoX Mayhem. We'd heard great things and user reviews are spectacular, but we just couldn't get the hang of it. We mangled our poor rider's carcass in some pretty unimaginable ways and came pretty close to mangling our iPhone in frustration too. But then, just as we were about to strike it from the list, everything clicked. Once we realized it was all about skill, patience and poise rather than speed, we weren't dying as often and the appeal became apparent.

GAMEPLAY

So, now that's out of the way, you'll probably want to know a little about the game itself, right? You're a motocross trial rider, challenged with making your way from one end of an island course to the other. The trouble is, the terrain is littered with jumps and perilous drops. If you hit them too quickly, too slowly or at the wrong angle, you're dead meat. Access the throttle by tapping and holding the right side of the screen, while the left side is the brake. Your rider can lean back and forward by tilting the phone, which helps him to balance when going uphill or downhill.

You will crash. Several times. Each time you do, you'll have to return to the start of the level and, annoyingly, each fall adds a second to your overall time. But it's a trial-and-error type of game. The key is to learn from the bone-shattering fall and apply the lesson on the next run. If you went at a jump too fast, then back off a little bit. Fell off the back of the bike? Lean forward next time. However, the crashes are truly magnificent. As frustrating as it can be, you'll find yourself chuckling at the way your rider crumples into a heap of broken bones, and the game is very easy on the eye in a 2D cartoony kind of way.

THE LOWDOWN

The free ad-supported game only gets you three levels and the ads are obtrusive. For a small fee, you can upgrade to the full version and continue the fun on multiple islands. There are even a couple of extension packs if you can't get enough motocross action.

You Might Also Like
Sticking with the motocross theme, Hardcore Dirt Bike is a decent racing simulator with really neat 3D graphics.

Hot Tip
Turn your rider into a bear by accessing the hidden Easter egg in the About screen.

NBA JAM

Category: Sports • **Available on:** iOS/Android • **Cost:** Paid
Download size: Medium • **Age rating:** Suitable for all

SYNOPSIS

He's heating up ... HE'S ON FIRE! Relive the best basketball game ever.

NBA Jam was a bona fide console classic back in the 1990s, and those who remember this title fondly or those who have yet to experience the wonders can now pick up this faithful adaptation on iOS and Android. Thankfully, it's still as enjoyable 20 years on. Just as before, this game takes the traditional basketball simulation, takes it down to two on two, and adds cartoonish and outlandish slam-dunks to exponentially add to the fun.

You can choose from any franchise in the NBA and pick your two favourite players from up-to-date rosters (apologies to those who don't speak American). You then have to battle your way through every team (they get progressively better) until you've downed them all, unlocking achievements and new players in the process.

GAMEPLAY

You control one member of the team using a virtual on-screen control pad that allows you to move around the court, pass, shoot, run and tackle. To perform a dunk, hold down the shoot button when running towards the basket and you'll be greeted with a 'BOOMSHAKALAKA' from the excitable commentator. When you don't have the ball, the buttons switch to sprint, jump and run. Hit jump to block a shot from your opponents, while hitting tackle can force a turnover. Your teammate is equally important. He's mostly controlled by the CPU, and moves, passes and shoots independently. You can override him, by making him pass or shoot, but don't dismiss him! He's a valuable ally, especially on defence!

When your player hits three baskets without reply, the commentators will announce, 'He's on fire,' which makes your dude superhuman. This is definitely the most satisfying part of the game, especially when you smash the glass!

THE LOWDOWN

The two-on-two format is what really makes this game. Scoring is faster, gameplay is simpler, games are more intense and the slightest changes in momentum are crucial. The one-player career mode is a blast, but you can take on a friend using the Bluetooth or Wi-Fi multiplayer modes. This allows four handsets to play with everyone controlling a player.

You Might Also Like
NBA: King of the Court 2 is a wicked arcade-style game where you progress by completing challenges with the aid of the device's motion sensor, while NBA 2K13 offers more traditional five-on-five action.

Hot Tip
To perform a wicked alley oop, pass the ball to your teammate and run towards the basket. Hold jump and he'll hit you in midair for an awesome dunk.

NEED FOR SPEED: MOST WANTED

Category: Racing • **Available on:** iOS/Android • **Cost:** Paid
Download size: Large • **Age rating:** Suitable for all

SYNOPSIS

Racing against a host of street-savvy speedsters is tough enough, but with cops looking to run you off the road, things can get tricky.

Avoiding the wrath of the fuzz while besting the other cars on the road is the challenge served up by this edition of Need For Speed. The result is an exceptionally playable thrill ride that amps up the adrenaline as you bid to become the Most Wanted racer in the game.

GAMEPLAY

The action takes place in the fictional city of Fairhaven, where you'll speed around great-looking tracks, unlocking over 40 cars (beat a car in a race and you unlock it) and earning virtual cash and experience points to spend on upgrades. There are various race modes – time trials, one-on-

one, no damage – and when you pick yours and get underway, the pedal is applied automatically to the metal. You can brake and reverse by tapping the right side of the screen, while tilting the device does the steering. Swipe up to activate your turbo and tap/hold the right of the screen to drift around corners.

However, the real skill (as all good criminals will tell you) is to avoid being caught. As you start each race, the police are waiting for you in numbers. They're littered along the course, trying to take you down as well as forming roadblocks. You'll need to avoid them or take them down. They'll try to ram you into walls and oncoming traffic. Ram them back without causing too much damage to your own car and you'll receive a turbo boost as a reward. Most satisfying. Damage is an element of this game, which differs from some arcade racers where your car is an indestructible pinball. As you hit walls and other road users during races, the damage will rack up on your vehicle and affect control and performance. Wreck completely and you're out of the race.

THE LOWDOWN

You can join EA's Origin platform to play online against fellow speedsters and you'll find loads of optional in-app purchases to boost performance, but there are hours of fun in the basic version of the game. Trust us. It took almost a day to write this page.

You Might Also Like
There are loads of games in this series, so try Shift or Hot Pursuit, but as much as we enjoyed Need For Speed: Most Wanted, we prefer Asphalt 7 (pages 136–37).

Hot Tip
Be on the lookout for shortcuts. The odd crafty diversion can push you up the field or further ahead of your rivals.

NEW STAR SOCCER

Category: Sports • **Available on:** iOS/Android • **Cost:** Free/Paid
Download size: Small • **Age rating:** Suitable for all

SYNOPSIS

The success, the fame, the cars, the money, the women. Live out your football dreams with New Star Soccer.

New Star Soccer is unlike anything you'll see in this book. It's part FIFA Soccer (*see* pages 144–45), part Football Manager (*see* pages 148–49) and part Flick Golf (*see* pages 146–47), and the result is one of the most addictive games to grace these pages.

You're a footballer starting his journey to superstardom, from the humble lower leagues all the way to his first cap for his country. You do this by completing passes, making tackles or making

shots on goal. Do well in a game and you'll earn rave reviews and your stock will rise with fans, teammates and management, and you'll earn sponsorships and receive offers from bigger teams. Miss chances and your opponents will capitalize. You'll be substituted, criticized and dropped.

GAMEPLAY

The gameplay is simple. To make a pass or take a shot, drag your finger backwards from the ball (the further you drag, the more power you generate), aim the arrow and release. The next screen controls the strike, allowing you to apply curl, height and extra power. The ball is often a moving target, so the real skill comes in timing. Practise in the Arcade mode. You can improve your skillset through training, and by buying better boots.

True to life, there's more to your success than assisting and scoring goals. There are lifestyle elements to consider too. You'll do media interviews and sometimes need to choose between fans and teammates. You can

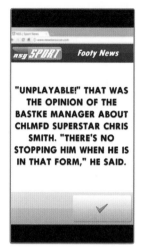

also buy clothes, cars and houses to increase your happiness, impress the ladies and net you a WAG. You can also gamble all of your cash away in the casino.

THE LOWDOWN

The free version of the game gives you a taster, but to play more than 10 matches, you'll need to unlock the full version. It costs less than a pound and it's well worth it. You can also use real money to boost your stats through the in-app purchase of Star Bux. Without these, the road to success is a long one, but the journey is half the fun.

You Might Also Like

Football Manager is the management equivalent of New Star Soccer, but for a true football simulator, you can't beat the FIFA Soccer series from EA Sports (pages 144–45).

Hot Tip

If your energy levels are lagging during an important game, take an energy drink at half time. It can make all the difference.

POOL BREAK

Category: Sports • **Available on**: Android/iOS • **Cost**: Free/Paid
Download size: Small • **Age rating**: Suitable for all

SYNOPSIS

You knew there had to be a pool game in here, right? This addictive simulation even squeezes in a spot of snooker too.

We enjoy nipping down to the local for a game of pool, but on those occasions when a night on the green baize isn't on the cards, we get our practice in at home with Pool Break. One of a few great titles available for both Android and iPhone users, Pool Break is a great-looking 3D game with realistic physics and an easy-to-master control system.

GAMEPLAY

You line up your shot by panning around the screen and pinching to zoom. Dragging the cue back shows where the cue ball will make contact with the object ball, which you can

adjust by moving left and right. You'll also see a yellow line depicting the trajectory of the object ball after contact (hopefully towards the pocket!). To increase the power, draw the cue back a little further and, once you're happy, swipe up to take the shot. You can also apply 'English': placing spin on the shot allows you to leave the cue ball in the perfect position for your next shot, which is a must for building breaks and keeping your opponent in his chair.

All of this is assisted by the game's lifelike physics. Experienced pool players will be able to apply their knowledge of the game to good effect, while novices will soon pick things up.

There are plenty of game modes to choose from, from traditional snooker to US 9-Ball, to good old-fashioned British eight-ball pool. You can even choose the table size, colour, shape (yes, shape!) and the environment in which the game is played. There's also a neat online multiplayer indicator when you're playing the computer. When this icon pops up, tap it to instantly start a game against a human opponent. You can also pass and play, using a single device to take on your friends.

THE LOWDOWN

The free app offers more than enough fun (there are also a couple of classic board games to play), but to get rid of the advertisements, unlock more game modes and play against Facebook friends, you can upgrade for a small fee.

You Might Also Like
There are loads of options here, but Miniclip's 8 Ball Pool is available for iPhone and Android and has a popular online multiplayer mode.

Hot Tip
To apply backspin on a shot, press the English button and strike the cue ball near the bottom.

REAL RACING 3

Category: Racing • **Available on:** Android/iOS • **Cost:** Free (with in-app purchases) • **Download size:** Large • **Age rating:** Suitable for all

SYNOPSIS

The clue is in the title. This game is so close to the real thing that you'll need a seatbelt, and the third instalment is the best yet.

Some of the other vehicle-centric games we've covered (like Asphalt 7 and Need For Speed: Most Wanted) have centred on arcade-style street racing, with turbo bonuses, weapons and huge crashes, while some are completely detached from reality (in the case of Zombie Highway, we hope so).

Real Racing 3 is different. It features actual circuits like Silverstone, Brands Hatch and many other familiar stretches of tarmac for you to master, while there are also 45 officially licensed cars from Porsche, Lamborghini, Dodge, Bugatti, Audi and more, all of which must be repaired if they're damaged.

GAMEPLAY

In terms of gameplay, racing novices get braking and acceleration assists, meaning your only concern is steering (tilting) at the right time. Once you become acquainted with the game, you can change the settings to manual. There are multiple modes to keep things interesting, like elimination, speed challenges and full 22-car grid races. Winning and performing races will earn you the in-game R$ and Gold currencies, while you'll also level up your driver, allowing you to progress to new races. The R$ can be spent on repairing and upgrading your car, while the Gold enables you to by-pass the time you have to wait for upgrades to complete. If you earn enough cash, you can buy new cars. This can take a while and a lot of racing.

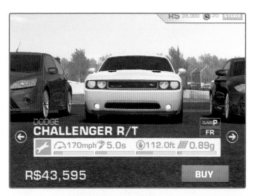

THE LOWDOWN

The game is free to download, but if you want to skip the grind of earning cars and cash on merit, then there are plenty of packs you can buy with real money to fast-track your driver and get you behind the wheels of the elite cars. This great-looking game also offers a neat Time Shifted Multiplayer innovation, which puts you against real people who have posted their times. This makes beating people way more satisfying and it's also a great way to battle friends offline.

Graphically, Real Racing 3 has more polish and shine than most mobile games. While the 'freemium' model isn't for everyone, we found plenty of enjoyment from just playing without spending real money.

You Might Also Like

If you're not a fan of the 'freemium' games and would rather pay a couple of quid to enjoy a full game, get Real Racing 2.

Hot Tip

Driving smart is as important as winning. Avoid collisions and harsh cornering. You want to spend your winnings on upgrades rather than repairs.

RECKLESS GETAWAY

Category: Racing • **Available on:** iOS/Android • **Cost:** Free/Paid
Download size: Medium • **Age rating:** Suitable for all

SYNOPSIS

Escape with the loot and take the cops down in this non-stop, action-packed thrill ride.

Why are so many of these driving games centred on criminal and evil deeds? Well, the truthful answer is, it's just more fun when you're on the wrong side of the law. In the devilishly enjoyable Reckless Getaway, your task is to successfully plough your way to freedom following a series of bank robberies. However, it's not the escaping, it's how you escape that really matters. You'll need to destroy the relentless cops on your tail, navigate oncoming traffic, run innocent road users into ditches, collect coins, make stylish jumps and collect neat power-ups to ensure it's a heroic escape.

GAMEPLAY

Once you take the wheel, the controls couldn't be simpler. You hit the right and left side of the screen to steer (tilt controls are available in the options), while also targeting point-scoring

power-ups, coins and ramps. You'll earn points for them all, with the idea being to obtain up to four stars for your getaway. Power-ups, within this visually enjoyable and accessible racer, include turbo boosts, the ability to jump items in your way (cars, police roadblocks and holes in the road) and the chance to daze your opponents, clearing the road temporarily.

Sounds easy, right? Wrong.

The police are homicidal maniacs who'll take themselves and anyone else down if it means halting your progress. They'll try to ram you off the road and destroy your vehicle. Although the high-speed crashes are hilarious, these endless collisions take a toll on your health bar and, if you wreck, the number of stars you can get for your escape is reduced. Deviating or being deviated from the escape route also costs you points.

THE LOWDOWN

You will always make it to freedom, but the idea is to cause as much destruction and earn as many points as possible without wrecking your own car. Tackle fellow road users? 250 points. Avoid oncoming traffic? 500 points. Jump over other cars? 500 points. Destroy a cop? 700 points. Achieve more than one at the same time and the tally is multiplied. The more stars you pick

up, the more frenetic Getaway levels you unlock. You'll encounter different terrains, tougher obstacles and more cops. You'll also have a blast.

You Might Also Like
Reckless Racing, also made by Polarbit, applies the same concept to an actual race.

Hot Tip
Find the right balance between destroying other cars and protecting your own vehicle. Wrecking your car costs you points and hence stars.

RIPTIDE GP

Category: Sports • **Available on:** iOS/Android • **Cost:** Paid
Download size: Small • **Age rating:** Suitable for all

SYNOPSIS

Leave your wetsuit at home and power your way to watery immortality on your lunch break.

The great thing about mobile gaming is that, at any point in the day, we can grab our phones and live out a different life for a few minutes. In the case of Riptide GP, we can become bad-ass, world-class jet-skiers. This tilt-to-turn arcade racer is just as much fun as you'd imagine it to be. It's a high-octane ride from start to finish and it looks absolutely fantastic too.

GAMEPLAY

Before the fun starts, you'll need to choose the arena, the difficulty level and the type of jet-ski you'd like to command. Gameplay is really straightforward. The auto-accelerator makes controlling your speed easy, while the occasional tap of the brake keeps your ski under control.

The real beauty of Riptide GP comes with the trick/reward system. You can perform a multitude of awesome flips and skill poses by swiping both thumbs across the screen using various direction combinations. For example, swipe outwards with both thumbs to perform a back flip, swipe down with both to fly like Superman. If you land the jumps, you'll be rewarded with a short-term speed boost that can be activated by tapping the meter on the right of the screen. If you wipe out in spectacular fashion (and yes, they are pretty spectacular), the arcade-style gaming means you won't die or drown as a result. All it takes is a tap on the screen to recover, but you'll have lost a few seconds on your competitors as a result.

THE LOWDOWN

As you progress in the championship mode, you'll unlock more courses and skis. There are 12 tracks, five championship modes and seven skis for you to work your way through. The action is exceptionally fast-paced, but the superb graphical performance never yields and the splashes of water on the screen when you land are a particularly nice touch.

The app costs a small fee, but after you've splashed out (sorry, we couldn't resist it), you can enjoy the full benefits of the game.

You Might Also Like
Shine Runner is made by the same company and applies the same gameplay principles to the noble art of bootleggin' through muddy backwaters.

Hot Tip
If you drive directly behind an opponent, you'll avoid the choppier waves he creates in his wake and have a smoother ride as a result.

SLINGSHOT RACING

Category: Racing • **Available on:** Android/iOS • **Cost:** Free/Paid
Download size: Small • **Age rating:** Suitable for all

SYNOPSIS

Slingshot Racing reinvents the wheel with a near-perfect, supremely original racing game.

Within this book, there's a little taste of everything. In this chapter, we've raced (and trashed) sports cars, trucks, speedboats, motorbikes and snowboards, but Slingshot Racing is pretty unique, and perhaps the most enjoyable, racing experience of them all. In this great-looking, top-down game, there's no sign of a steering wheel, accelerator or brake pedal. Instead, vehicles are controlled by latching on to grappling hooks and slingshotting around corners, with the aim of completing the weird and wonderful courses faster than human or computer-controlled opponents.

GAMEPLAY

It's all done with one-touch control too. You press and hold one corner of the screen to attach a cable and release your grip to detach. Hold again to grab the next cable. Sounds easy, right? Well it's not. Slingshot Racing is all about timing and momentum, so you'll need to grab and release cables at the right time to maintain speed, keep your car on the racing line, avoid hitting the walls and also grab speed bonuses.

If you play the career mode, you'll unlock new Tours as you progress, with up to 80 races to take part in, each more inventively designed than the last. There's also the incentive of earning the full three stars for each race you complete. Not that you'll need any encouragement – this is dangerously addictive. Sound good so far? We haven't even got to the best feature yet.

The local multiplayer mode allows up to four people to play against each other on the same device. How is that possible? Well, each player is allocated a corner of the screen to control a colour-coded vehicle and away you all go. Suffice to say it's easier on a tablet than having everyone fighting over a phone, but it's a really unique feature that single-handedly secured Slingshot Racing's spot in this book.

THE LOWDOWN

In the iOS App Store you can grab Slingshot Trials, which is a free version of the game, allowing you to try before you buy, but trust us, this will be money well spent.

You Might Also Like
This game has no peers, making it hard to recommend an alternative. It's not a racer, but ball-rolling game Gears is also made by developer Crescent Moon and is also awesome.

Hot Tip
If you're struggling to keep up with your opponents, keep a close eye on where they're grabbing and detaching cables and do what they do.

STEAMPUNK RACING 3D

Category: Racing • **Available on:** iOS/Android • **Cost:** Free (with in-app purchases) • **Download size:** Small • **Age rating:** 12+

SYNOPSIS

Enter the savage world of steam-powered death machines with this brutal racer.

Think Mario Kart set in a war-torn, post-apocalyptic industrial world, but without the endearing characters, and you'll already be most of the way towards understanding what Steampunk Racing is all about. The idea is not just to win races in your makeshift vehicle, but to completely destroy the opposition in the process in an all-out battle to the death. The only problem is, they're trying to do the same to you and they're pretty darned good at it.

GAMEPLAY

Steampunk Racing 3D is another tilt-to-turn arcade-style racer, but it's the strategic laying of traps and unleashing of weapons that set this game apart from the others. As you drive over

steampiles on the track (fellow racers can grab them too), you'll fuel your arsenal, which differs depending on the vehicle you choose to lead into battle.

You can leave spikes or oil slicks for those behind you, unleash blades to take down rivals alongside you, send out EMP waves to kill your opponents' energy levels and even teleport ahead of your enemies. The steam can also be used to give your death machine a much-needed speed boost. The weapons look great, but only slow down your rivals temporarily, so you'll need to keep hammering them with gusto.

As you complete the various missions, you'll earn coins, which can be spent on new vehicles, unlocking more tracks and improving the capabilities and weaponry on offer. To seriously progress, you may want to buy more coins using in-app purchases, but there's plenty of enjoyment to be had in the free version of the game.

THE LOWDOWN

Beyond the single-player campaigns, you can also take your penchant for vehicular assault to the internet, where you can play online against other kill-crazy fiends. There's a quick-play mode, where you face off against a random foe or you can enter online tournaments, but they'll cost you a ton of coins.

The game is probably better played on a tablet, because the controls can be quite fiddly, and the text is quite small and difficult to decipher when using smartphones.

You Might Also Like
Road Warrior Racing is a fun 2D alternative. You perform flicks and tricks while taking down your opponents with weapons.

Hot Tip
Enable Auto-accelerate in the options menu. It simplifies gameplay and leaves your hands free to fire your arsenal of weapons.

STICK CRICKET

Category: Sports • **Available on:** iOS • **Cost:** Free/Paid
Download size: Small • **Age rating:** Suitable for all

SYNOPSIS

A simplistic but endlessly fun cricket game that hits the competition for six.

Within this book, there aren't too many games that are only available on Apple devices, but we had so much fun playing Stick Cricket that we just had to include it.

The game channels the spirit of the Twenty20 form of the game, with the idea being to hit as many runs as you can in a limited number of overs. Refined it is not. Despite the stick-figure graphics and simple two-button gameplay, this is the best cricket game available on mobile devices.

GAMEPLAY

It's played thusly. Your computer-controlled opponent bowls to you and you hit the left or right buttons, depending on the ball's trajectory. Where's the skill in that? Well, if you don't time it, you're going to be skittled out, caught by fielders or out LBW. If you miss a bouncer, it'll knock you and your stumps out.

You'll face fast, slow and spin bowlers and the timing required differs for each. Some deliveries come down the middle of the track, so you'll need a quick decision over which way you'll swing the bat. Go the wrong way and your well-animated stick man will be on his way back to the pavilion.

There are a couple of different game modes. All-Star Slog encourages you to score as many runs as possible in five overs, with a fully licensed international team playing against a team of bowling all-stars from the past. That's a lot of fun, but the World Domination mode is more challenging and ultimately more rewarding. You take a team of legends (such as Botham, Lara, Ponting) and have 20 overs to chase down increasingly tougher scores from increasingly difficult bowling attacks. It's exhilarating and if you can conserve wickets you'll have a chance when the big slog comes in the final few overs. Getting the win unlocks the next team in your path.

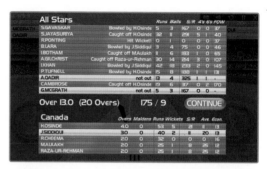

THE LOWDOWN

The game is ad-funded, so it's free for iPhone but costs a couple of quid for the HD iPad version. You can unlock more teams and play longer All-Star Slog matches through in-app purchases.

This game looks great and plays well, and it's definitely one to kill some time on those days when rain stops play.

You Might Also Like
For more advanced graphics and gameplay, try Cricket World Cup Fever, while Stick Tennis applies the same principles to an equally enjoyable sport.

Hot Tip
When facing fast bowlers, take your swing just before the ball bounces.

TIGER WOODS PGA TOUR 12

Category: Sports • **Available on:** iOS/Android • **Cost:** Paid
Download size: Medium • **Age rating:** Suitable for all

SYNOPSIS

Take Tiger's crown and dominate the tour in this ultra-realistic golf game.

Thanks to huge advances in mobile technology, full console-quality games are now available to play on Android and iOS devices. The classic Tiger Woods PGA Tour franchise from EA Sports is one of those titles.

Now, Tiger himself may have slipped a bit in recent years, but the game that bears his name just keeps getting better. This is clearly the most detailed, best-looking and most comprehensive golf game you can buy for your smartphone or tablet.

GAMEPLAY

The idea, of course, is to get the ball into the hole in as few shots as possible, but the key gameplay principle is to replicate a smooth golf swing. Take a backswing by pulling down on the power meter and (without letting go of the screen) flick up to follow through. The distance/accuracy of the stroke will depend on the smoothness and straightness of the motion. Once the ball has left the club, you can apply multidirectional spin by furiously swiping the ball. If you feel like you've slightly overhit the shot, drag down to reel it back.

Sounds easy? Well there are sand and water hazards and rough grass to contend with on these beautifully reimagined real-life courses, and you have to make sure you select the right club, otherwise you'll fall short. When you get on the putting green, things get even trickier. But there's help on hand. The caddy offers some good tips on the pace and undulations of the green, and there's a neat putt-preview mode to simulate how the putt you've lined up will travel.

THE LOWDOWN

There are a number of game modes. You can play as Tiger (or a number of other real-life golf pros) in exhibition games, but the career mode is where the challenge truly lies. You take your own golfer on tour and look to earn money by competing in full

tournaments against the best. The better you perform, the more money you earn, which can then be spent on upgrading your golfer's skillset. You can also earn money by performing mini Tiger challenges, which enable you to pay the entry fee for more tour events.

You Might Also Like
Tiger is the most detailed and realistic golf simulator, but for a pick-up-put-down experience, try Flick Golf (pages 146–47).

Hot Tip
For a power boost, stay on the shot until the meter turns red and then flick up to follow through.

VIRTUA TENNIS CHALLENGE

Category: Sports • **Available on:** iOS/Android • **Cost:** Free/Paid
Download size: Medium • **Age rating:** Suitable for all

SYNOPSIS

Great controls, fabulous graphics, plenty of tournaments and fast-paced gameplay. What's not to love?

Virtua Tennis Challenge sees the popular console simulator smash its way on to handheld devices and serve up an ace in the process. You take on a variety of opponents on different surfaces in different tournaments with the aim of rising to the top of the world rankings.

GAMEPLAY

Part of the beauty of this graphically superior game is the intuitive touchscreen control system that requires you to draw your shot trajectory on the display. Let's start with serving. Just tap and release the display and swipe in your chosen direction. When the rally gets going, swiping up adds top spin to a shot, while swiping down allows your player to counter a powerful return with a slice shot. You can also adjust the length of your shot, with shorter swipes or play lobs and drop shots.

Naturally, shot positioning is controlled by the direction of the swipe, while the power of your stroke is determined by exquisite timing. The idea is to play your stroke as soon as the ball leaves your opponent's racquet. As soon as your stroke is played, you can tap the screen to reposition your player. For example, if you've hit a great shot that may be tough to get back, tap to come into the net. If you're expecting a rocket return, drop back behind the baseline.

Successful mastery of the controls will soon result in tournament wins and ranking points, earning you the prize money necessary to enter bigger events. The early opponents are easy to take down, but as you progress, prepare for long rallies and savour in the joys of out-thinking and wrong-footing the best in the world.

THE LOWDOWN

Beyond the single-player career mode, the multiplayer aspect allows you to take on friends over Bluetooth, while you can also battle random human foes over Wi-Fi.

The free version of the game is more like a training camp that allows you to get used to the controls and learn all of the shots. If you're taken in, then the full title won't cost you much.

You Might Also Like

If you liked Stick Cricket (pages 176–77), the chances are you'll like Stick Tennis. It's much simpler than Virtua Tennis Challenge, but still plenty of fun.

Hot Tip

Building the concentration meter allows you to pull off a Super Shot to match your player's style (for example, Atomic Forehand). When the meter is full, swipe with two fingers and make it count.

VIRTUAL TABLE TENNIS 3

Category: Sports • **Available on:** iOS/Android • **Cost:** Free (with in-app purchases)
Download size: Small • **Age rating:** Suitable for all

SYNOPSIS

Become a ping-pong master with a swift flick of your finger.

Table tennis, or ping pong for those with a preference for fun over stiff competition, is an almost universally enjoyable pastime, so it's little surprise that it features among the best sports games available on Android and iOS today. Virtual Table Tennis 3 belongs to the flick-to-hit

school of touchscreen sports titles. You manoeuvre your paddle around the screen and swipe upwards with the requisite speed and direction, with the aim of making it as difficult as possible for your opponent to get the ball back in play.

GAMEPLAY

We must stress that it's easy to get disheartened by playing this game in the early stages. There's no real tutorial and you're expected just to pick it up. However, stick with it and you're in for some real fun. The key, we found, is to keep a finger or thumb on the bat at all times, almost obscuring it from view. When the ball comes your way, move towards it and flick to return it to the other side of the table. The longer your flick, the harder you hit the ball. For more power, keep your paddle at the back of the court and swipe through the ball. The direction of your swipe will help you keep it in the corners, where your opponent will find it harder to return. When you've set the point up to your satisfaction, you can finish it off by flicking to smash.

THE LOWDOWN

There are several one-player game modes and difficulty settings that allow you to compete in single arcade-style matches against a host of opponents of varying skill. You can also compete in tournaments, earning points for every win, which you can spend on improving your paddle and increasing your energy levels.

The real fun, however, comes with the multiplayer modes, where you tackle a real opponent rather than the computer. You can hook up to another phone via Bluetooth, or play someone online over your Wi-Fi connection. Remember, winning is far more satisfying when you can rub it in someone's face.

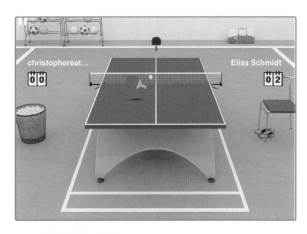

You Might Also Like

World Cup Table Tennis is good fun, but it's simply not as good. Virtual Table Tennis developer Clapfoot also makes a tennis equivalent called Play Tennis.

Hot Tip

To fool your opponents, you can apply side spin to your shots by swiping perpendicularly to the direction of the ball as you swing your bat.

ZOMBIE HIGHWAY

Category: Racing • **Available on:** iOS/Android • **Cost:** Free/Paid
Download size: Small • **Age rating:** Mature

SYNOPSIS

Take down as many zombies as you can on your way to certain death.

Zombie Highway isn't strictly a race; it's more of an endless highway to hell, where you drive until the undead flip your car and eat you alive. Your death is inevitable; it's just a case of how long you can survive. If that sounds terrifying, that's because it's meant to be.

GAMEPLAY

You take the wheel on a post-apocalyptic stretch of road, where the only traffic is the derelict cars of the fallen and the mindless zombies with a hunger for your brains. As you drive along the road – steering only by tilting your handset left and right – zombies will leap on to your car and attempt to flip you over. To avoid this fate, you can steer out of their way or turn them into roadkill. If they manage to hold on, the only ways to knock them off are to tactically glance other cars or to shoot them off by tapping the red circles

at the side of the screen. The more zombies on your car, the harder it is to steer. You then run the risk of hitting too much of a car while trying to dislodge them. If you do that, you're dead. If the zombies flip you, you're dead.

As you get further into the run (and it takes a while to get anywhere!), you'll come up against tougher zombies, who can regenerate, and fatter zombies, who'll use their weight to pull you over, but you'll also unlock more powerful weapons to take them down faster.

THE LOWDOWN

Make sure you've got the volume up, or headphones on, because the sound effects really enhance the experience. The roars of the zombies as they leap towards you may be unsettling, but the splat when you kill them is very satisfying.

There's a host of other, more zombie-resilient cars and new game modes to explore, but to access those you'll need to

pay a small fee to upgrade to the full version of the game. However, there's plenty to keep you occupied in the free version for the first few days at least.

You Might Also Like
If killing zombies is your thing, then give first-person shooter Call of Duty: Black Ops Zombies a shot. If you'd rather not battle them up close, take them out from the skies in Zombie Gunship.

Hot Tip
If possible, conserve your ammunition and use the debris to dispose of zombies instead. Also, a quick turn as the zombie leaps could make him miss your car.

SIMULATION & STRATEGY GAMES

If you're looking for a challenge or just want a game that you can get your teeth into, then you've come to the right place. In this chapter is a selection of games that will have you waging wars, building businesses and spending time with TV's most famous family.

A phrase you will come across in this chapter is 'tower-defence game'. This is a genre that has sprung up in the last decade and has thrived on mobile platforms. It basically does what the name

Above: Jelly Defense is a cute and rather fun take on the tower-defence genre.

suggests: you must defend your position with a series of towers, strategically positioned, from marauding enemies. It's been argued that the granddaddy of tower defence is an old game called Rampart that was around in the early 1990s. The genre has come on in leaps and bounds since then and this chapter will show you the very best it has to offer.

FUN AND FLEXIBLE

We're seeing developers enhance the genre by subverting it, as is the case with Anomaly Warzone Earth, or making it a fun experience for all, as with Jelly Defense, or even just making it as flexible as possible, as you'll see in Guns 'n' Glory. We loved them all and we're sure you will too.

Also in this chapter, we have games where you can build, shape and create your own world. There's a wide variety of worlds on offer, from the primordial soup of Jurassic Park to the familiar, brash world of the Simpsons. These games allow you to build familiar landmarks

Above: The intentionally low-quality graphics of Minecraft encourage your imagination to run wild.

from these popular franchises and give them your own spin.

FIND YOUR INNER NERD

If you want something that's more flexible, then you won't need to look any further than Minecraft. The game that's taken home computers by storm is now available on mobile platforms. Its intentionally low-quality graphics hide a game that speaks to your inner nerd and it's a game that encourages your imagination to run wild.

It would be negligent of us not to mention Plants vs. Zombies. It's already wildly successful, but if you want to see a title that's at the top of its game, it's this one.

BE AN EMPIRE BUILDER

Also look out for games like Restaurant Story and Zoo Story, where you run your own business and build it up into an empire. Of course we also have games that let you run an actual empire, such as Majesty or Little Empire.

Whether you're defending your kingdom's honour or just walking your virtual dog in The Sims, we've got you covered with the very best simulation and strategy games out there.

Above: The Sims is an immensely popular simulation game, in which you create virtual characters.

JULIAN'S PICKS

We hope you enjoy playing these games as much as we did. For the record, here's my personal Simulation and Strategy Top 10:

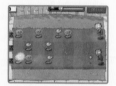

❶ PLANTS VS. ZOMBIES

(*see* pages 228–29)

There's a zombie apocalypse going on and mankind's best defence is ... plants?

❷ ANOMALY WARZONE EARTH HD (*see* pages 192–93)

Movie-level production qualities abound as you protect Earth from an alien invasion force.

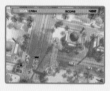

❸ PLAGUE INC. (*see* pages 226–27)

Star in your very own disaster movie, except this time, instead of saving humanity, you're trying to make it extinct.

❹ THE SIMPSONS: TAPPED OUT (*see* pages 234–35)

Homer's caused a meltdown (again!) and it's up to you to rebuild Springfield with the aid of all of your favourite Simpsons characters.

❺ EUFLORIA HD *(see pages 200–01)*

One of the most intriguing games you'll ever play asks you to save a colony from a dangerous foe.

❻ JELLY DEFENSE *(see pages 214–15)*

Defend your green crystals from the jelly invaders in this cute tower-defence game.

❼ MINECRAFT *(see pages 222–23)*

The home computer smash hit is now available on mobile devices. Be prepared to expand your world.

❽ GUNS 'N' GLORY *(see pages 212–13)*

Tower defence gets a new twist as you relieve settlers of their cash and worldly goods in the Old West.

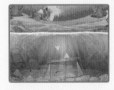

❾ EAGER BEAVER *(see pages 198–99)*

Edgar the beaver is a dam perfectionist and only he can save his village from the rising waters of the river.

❿ FRAGGER *(see pages 204–05)*

A game where throwing grenades at helpless individuals is perfectly acceptable.

ANOMALY WARZONE EARTH HD

Category: Strategy • **Available on:** Android/iOS • **Cost:** Paid
Download size: Medium • **Age rating:** Suitable for all

SYNOPSIS

Movie-level production qualities abound as you protect Earth from an alien invasion force.

Anomaly Warzone Earth HD might be like nothing you've ever seen before. We say this for two reasons: one is that the production levels on this game are way beyond what you think a game on a mobile platform would be capable of; and two, because this isn't your ordinary tower-defence game.

Normally if you're playing a tower-defence game, you're asked to defend a position on your map from invading troops. That mechanic is in play here, but instead of being the side placing the towers and traps, you're the side that has to overcome them. It's a fantastic idea and one we haven't seen used enough. If your competition is Anomaly Warzone Earth, however, we can see why people haven't tried!

GAMEPLAY

As soon as you start the game, you're faced with a fantastic introduction video that sets the scene. It looks amazing and in HD it takes your breath away. Once you get into the game proper, there are even more visual flourishes.

You start off in tactical view as you plan your way through the streets, trying to avoid the enemy's defences. You will encounter them, but you're best off trying to pick the less battle-intensive route. Once that's done, switch to battle mode and prepare yourself. Battle mode is viewed top down and the detail involved is amazing as you make your way through rubble-strewn streets and downed planes. It's the most beautiful warzone you'll ever see.

You're not completely defenceless, as you're able to collect power-ups to repair your vehicles, lay down cover or provide a decoy target. The enemy may be strong, but you've got enough tricks to hold your own.

THE LOWDOWN

The soundtrack is terrific and there's a great voiceover during the mission briefings that really adds to the cinematic atmosphere. It makes the whole game feel like an action movie, in which you're the star. In short, this game totally blew us away. Production values on this are sky high and we have no hesitation saying this is one of the best-presented games available right now.

You Might Also Like
Take the battle to the next level with Anomaly Korea.

Hot Tip
When repairing your vehicles, make sure you place the power-up slightly ahead of them so they can drive through it and your convoy gets maximum help.

ANT RAID

Category: Strategy • **Available on:** Android/iOS • **Cost:** Paid
Download size: Medium • **Age rating:** Suitable for all

SYNOPSIS

It's a bug's life as you defend your ant colony from the angry insects in your garden.

On the surface, Ant Raid looks like a fairly easy, superficial game. Underneath, it's a match for anyone who has flexed their muscles in the strategy game arena. Ant Raid sees you controlling a colony of ants. The colony changes to different locations, but the premise is always the same: you must protect your home from invaders in the form of other garden insects. We're not sure why they're so angry with the ants, but rest assured, the ants are the good guys.

GAMEPLAY

The ants form a defensive perimeter around the colony and you protect it by deploying your fearless ants to attack the enemies. This is done in one of two ways: either you can tap a section of ants or you can press the screen and keep holding to select a wide circle of ants.

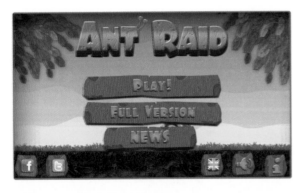

Once you've selected the required number of ants, you deploy them in the direction you want them to go, using the on-screen arrow that appears.

It sounds simple and, in the early levels, that double as tuition levels, it is indeed straightforward. However, bigger things are on the horizon, bigger enemies, faster enemies, enemies with an array of different attributes. All of which means that rather than sending your ants off gung-ho, you need to start planning tactically for all the oncoming enemies. In addition, your wounded ants need to be recovered by medics to be taken back to the colony so they're ready for action again – the brave little heroes.

Ant Raid has a star scoring system and a survival mode should you fancy a change from the main story levels, of which there are well over 90. There are also power-ups to look for during combat, which will give your lads the edge when taking on the bigger enemies.

THE LOWDOWN

The presentation is done with eye-catching 3D visuals and the different gardens and colonies are nice and varied. We especially liked the music with its chorus of singing ants and the little comments the ants make during combat always raise a smile. The story is told with little ant pop-up bubbles and the whole game is a very neat and pleasing package.

You Might Also Like
If you're after something by the makers of Ant Raid, try Farm Frenzy, in which you have to look after your own farm.

Hot Tip
Don't send every ant to tackle a giant snail. They explode, hurting your ants in the process and leaving you vulnerable.

DEFENDER 2

Category: Strategy • **Available on:** Android/iOS • **Requires:** iOS 4.3+/Android 2.0+
Cost: Free (with in-app purchases) • **Download size:** Small • **Age rating:** 9+

SYNOPSIS

Dragons! Goblins! Err... eyeballs! Defend your castle!

It's difficult to categorize what sort of game Defender 2 is. You're in a tower and waves of enemies come out at you, but it's not your normal tower-defence game. It's more like a survival game, or maybe wave defence.

GAMEPLAY

You're a hero in a world of monsters, goblins and all sorts of foul beasties. You have to defend your castle from said beasties with your trusty crossbow. You do this by simply tapping on the oncoming enemy to fire your crossbow. Obviously, enemies don't go down with one hit, so you can either continue tapping at the enemy as they advance or simply hold the screen. Alongside your trusty crossbow, you have some magic spells (naturally). They're kept at the bottom of the screen and when you want to unleash your fury upon your enemy, just drag the spell over the appropriate area of the screen and all your monster-related troubles are dealt with.

Defender 2 possesses an intricate levelling-up system whereby you earn points and coins for every monster you've slain and depending on the condition of your castle. When you've gathered enough points, you'll advance to the next level, with each level granting you access to new spells and abilities.

Coins work in a similar way, as you buy upgrades for your weapons and fortify your walls against the oncoming attack. Other items in the game can be purchased with crystals. You have a few to start off with, and you earn more throughout the game, but if you want many more, you can buy them at the respective app stores.

THE LOWDOWN

The enemies are nicely designed (although we did recognize the floating eyeballs from another game franchise) and the background is pretty enough. We loved the music, which, whilst nothing new, seemed to hark back to the music you used to find in old arcade games of the late 1980s.

Defender 2 is a great way to pass the time. The role-playing elements on the upgrade system are especially welcome, as is the online play where you and a friend can face the hordes together and see who lasts longer. In terms of value, very few games can claim to have a better offer for you than Defender 2.

You Might Also Like
The developer, DroidHen, obviously likes the mythical-creature genre. Try Fort Conquer for no other reason than its a brilliant name.

Hot Tip
Use the online battle mode to gain more coins for your single-player campaign.

EAGER BEAVER

Category: Strategy • **Available on:** Android/iOS • **Cost:** Paid
Download size: Small • **Age rating:** Suitable for all

SYNOPSIS

Edgar the beaver is a dam perfectionist and only he can save his village from the rising waters of the river.

Eager Beaver is a game that looks easy to play but is in fact a real challenge for anyone. In theory, dropping blocks into a river to dam it up should be easy, but Eager Beaver has a sophisticated water design that makes a simple task extremely difficult.

GAMEPLAY

Edgar the beaver needs to build dams and quickly the water level of the river is rising and none of his beaver friends will help him. They say it's because he's so good at it that he doesn't need their help, but we've tried that excuse before and it's never worked for us. Edgar has a crane and he has different-shaped rocks; you have to drop the rocks into the water, building them up to form a dam. Later on, Edgar will use an assortment of other items such as oilcans and, our favourite, old car licence plates.

It's first and foremost a case of matching shapes to make the best fit, but where the game really excels, and where the challenge comes in, is the water itself. It's the drift of the object that you have to factor in when making your dam. You might have the rock lined up on the crane correctly, but you need to account for the drift of the water when you drop it. The rocks rotate on the crane, so it's as much a case of timing as anything else. Control is a simple one touch for the crane as you drag it back and forth before releasing the rock.

You also have to look out for obstacles such as mines and fish along the way. They're all a threat to your dam, although we did find the mines useful once or twice, though we're not sure that's what they were intended for.

THE LOWDOWN

There are lots of villages on offer and there's the star rating system that ranks how well you've built your dam. The animation is wonderfully smooth and the design of Edgar and his surroundings are very cute: look out for the frozen caves on level four, where Edgar dons a lumberjack shirt and trapper hat.

We found Eager Beaver to be a real joy to play, so don't be fooled by its playful exterior, because this game is tougher than a beaver's front teeth.

You Might Also Like
Amazing Alex (pages 80–81) from the makers of Angry Birds will keep you entertained for hours with similar puzzling action.

Hot Tip
Don't let the fish out of their cages!

EUFLORIA HD

Category: Strategy • **Available on**: Android/iOS • **Cost**: Paid
Download size: Medium • **Age rating**: Suitable for all

SYNOPSIS

One of the most intriguing games you'll ever play asks you to save a colony from a dangerous foe.

Never before have we encountered a game that is so committed to getting you to relax as Eufloria HD. During the early levels, the game tells you can speed up the action if you wish, then it turns around and says that if things get too hectic then just slow down and enjoy the game. When you start playing, the game also asks you to plug your headphones in, and we can understand why. The musical accompaniment here is like something you'd listen to whilst meditating or getting a massage; it's almost blissful.

GAMEPLAY

At its heart, Eufloria HD is a real-time strategy game. The game's design is very minimalist, which makes it look sparse, but beautifully so. When the game talks to you, it does so through the guise of the mother tree, its words soothing and laid back. It's quite a surreal experience.

The game's goal is for you to colonize other asteroids with your seedlings. You do this by sending them from asteroid to asteroid. When you send 10 seedlings or more to an asteroid, you can colonize it with a tree. The tree creates more seedlings, up to a total of

40. At that point, it stops producing until you move some of the seedlings to a different asteroid. This is done with a one-touch method, which is extremely easy to use. When you want to move to a new asteroid, you drag an arrow from your current base. A green circle appears and you have the option to send as many seedlings as you want, but the best option is to send a scout ahead.

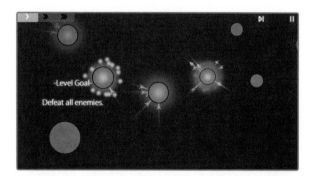

The enemies in Eufloria HD are the grey seedlings. They're aggressive, want to attack you and will put up a fight if you try to colonize their asteroids. Once you've established how many there are, you can send your forces across for your own version of aggressive expansion.

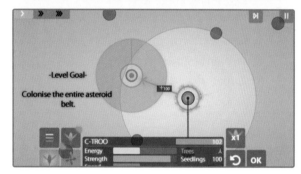

THE LOWDOWN

There are 25 levels, plus additional modes that test your skills further. Eufloria HD isn't a game that's interested in getting your pulse quickened, it's far more interested in getting you to relax and conquer asteroids at a leisurely pace. It's a sentiment that appeals to the laid-back hero in all of us.

You Might Also Like
Try Osmos HD, which has a similar design style to Eufloria HD.

Hot Tip
If you want to make scouts faster, plant two trees on an asteroid.

FIELDRUNNERS

Category: Strategy • **Available on:** Android/iOS • **Cost:** Paid
Download size: Small • **Age rating:** Suitable for all

SYNOPSIS

Defend your ground against an overwhelming number of enemies. They have tanks, but you have guns... big guns.

Fieldrunners is a tower-defence game that takes no prisoners. Its visual style might defer to a cartoon aesthetic, but the game itself is tough, requires you to strategically plan where you're putting your defences, and it doesn't let up.

GAMEPLAY

There are eight different types of tower, but you can only use four types on any one map. These towers range from the basic Gatling gun, which mows down the weaker enemies, up to the lightning tower, which can polish off most enemies with a single zap. All of these towers can be upgraded to make them even more fearsome. The skill with Fieldrunners is in making the path to your base as convoluted and maze-like as possible. The enemy need to run round in circles so you can turn them into mulch.

In action this frantic, the menus need to be simple, and they are. All of your towers are laid out at the bottom of the screen and all you need to do is drag your tower into the desired spot and that's it. Destroying each enemy earns you money, which is displayed at the top of the screen, and you use this extra income to buy more towers. There are three modes available: classic, extended and endless. Both classic and extended limit the number of enemy attacks thrown at you, and endless, well, you can pretty much guess. There's a save-game function, so don't worry, you won't have to end the game prematurely if you have other things to get on with.

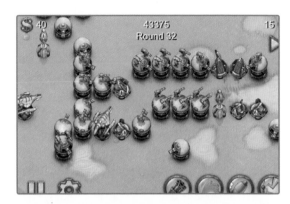

THE LOWDOWN

The cartoony style lends itself well to the game, which has a rather irreverent edge to it. Which is probably for the best because if it were played straight, it would make pretty grim viewing. Also enhancing the experience, the music never gets boring, even when you're playing in endless mode.

Two levels are unlocked and a further two can be unlocked as you progress. That's more than enough to be going on with, but, if you think you're up to the task, there are also four levels that you can buy should you choose. It's a challenge however you take it on!

You Might Also Like
Anomaly Warzone Earth HD (pages 192–93) takes the tower-defence genre and flips it on its head. Tremendous fun.

Hot Tip
Use the goo towers in the middle of your Gatling guns to slow your enemies down and cause maximum carnage.

FRAGGER

Category: Strategy • **Available on:** Android/iOS • **Cost:** Free
Download size: Small • **Age rating:** 9+

SYNOPSIS

A game where throwing grenades at helpless individuals is perfectly acceptable.

We're not sure who Fragger is, but we'd definitely like to know his motivation. He loves tossing grenades at people. We're not told who these people are, but we'll just assume that they're bad guys. What we do know is that Fragger's job is to lob grenades at other people dotted around the screen. Well, we say job, but it could be more of a hobby, as he clearly enjoys doing it.

GAMEPLAY

You throw the grenades by holding your finger on the screen and dragging the on-screen arrow up and down in the direction you want to throw. With your finger still pressed on the screen, you will need to slide left to right to adjust the power. It takes a little getting used to at the beginning, but after a few levels it feels very natural. Once you've established your trajectory

and power, you release and your grenade goes sailing away. If it lands with precision, you'll take out the enemy. If you've misjudged, it'll go flying away and be no good to anyone.

The whole setup is very simple and you should coast through the early levels. There are currently 12 different stages, with each containing at least 30 levels, a staggering number for a free game. If you get stuck, you can ask the game to show you the solution or allow you to skip the level. However, you only have a finite number of these, and to get more passes you need to purchase them.

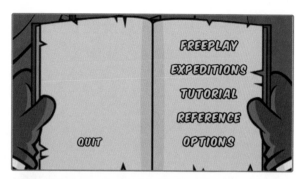

THE LOWDOWN

The animation is very cute, with Fragger getting different costumes for each stage. This doesn't add anything to the gameplay, but it's fun to see him going about his business in a Spartan costume or dressed like the Easter Bunny.

There's enough in here to keep you entertained for hours and it's a surprisingly bulky package for a free game. The game gets increasingly fraught as you move through the levels, and the tactical element comes into play later on as you try to move objects out of the way or navigate your grenades through small areas. It all adds up to a must-have game that's simple to play and easy to lose yourself in.

You Might Also Like
Miniclip have another game called Anger of Stick 2, which is also comically violent.

Hot Tip
Don't use too much power on long throws. Let gravity help you out, especially in tight spaces.

GEM MINER 2

Category: Strategy • **Available on:** Android • **Cost:** Paid
Download size: Small • **Age rating:** Suitable for all

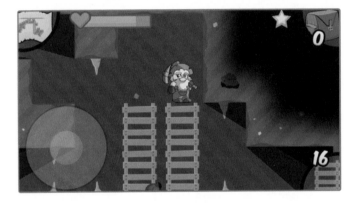

SYNOPSIS

Go underground to dig up precious ore, minerals and rare artefacts. It's a dangerous job, but someone's got to do it.

Gem Miner 2 tasks you with delving deep into the earth looking for metals and minerals. You're going in alone and you've only got a pickaxe for company. But even if you're claustrophobic or afraid of the dark, it doesn't really matter, as Gem Miner 2 is a very friendly and downright addictive game. The first thing you should do with Gem Miner 2 is get into the tutorial. The game isn't difficult to understand, but we've yet to see a better or clearer tutorial than the one on offer here. Once you've completed that, it's off to mine some gems.

GAMEPLAY

There are two different modes of play – freeplay and expeditions. Freeplay places you in a mine, and from there you're able to do whatever you want, be it getting gold or other minerals for money, or hunting down precious ancient artefacts. Expedition mode has a tighter mission structure, where you're tasked with finding a specific number of certain items. In each mode, you can leave the mine and visit the shop on the surface. In here, you can buy items to help you underground.

You control your miner using an on-screen joystick, which is also used to dig. Digging is also easy: you just hold the joystick down in the direction you want to dig. If all this sounds a bit too easy, then there is this to consider: you have a stamina bar that decreases when you climb ladders and mine things, so you need to keep that topped up by purchasing sandwiches at

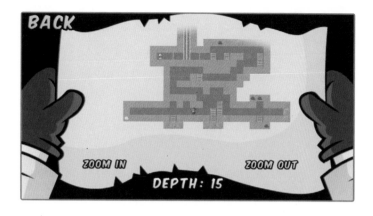

the shop or resting on the surface. You also carry items like dynamite and ladders. If you run out of ladders or get lost underground, you have to call for help. This will put you back on the surface, but you'll lose everything you've collected. There are also fires, gas pockets and water further hampering your progress.

THE LOWDOWN

The graphics are big, chunky and clearly defined. There's no music, but the sound effects are good and don't get in the way. It's a very easy game to play, but there's a real challenge here, between searching for minerals and managing your inventory, making sure you can stay underground for the longest possible time. This is one game that deserves to see the light of day.

You Might Also Like
Try the original Gem Miner and Gem Miner: Dig Deeper if you want more digging action.

Hot Tip
Watch out for the white rocks; they collapse easily.

GHOSTBUSTERS

Category: Strategy • **Available on:** iOS • **Cost:** Free (with in-app purchases)
Download size: Medium • **Age rating:** 9+

SYNOPSIS

The Ghostbusters are back in a game that asks you to save New York from another supernatural threat. Remember, no job is too big, no fee is too big.

There's a tower in New York that's got some weird swirly thing (technical term) above it and you have to get to the top. The tower's 50 floors high and has slime blocking the door on each level, which means you have to go around New York taking on other busting jobs and collecting slime samples to get to the top. In reality, this is the game asking you to level up, but it's well disguised and integrated nicely into the Ghostbusters universe.

GAMEPLAY

Your Ghostbusters are all rookies, so their busting abilities, equipment and shield defences all need to be upgraded as the game goes along. You can do this by busting ghosts, purchasing

new items in the shop, or from pick-ups found whilst out on jobs. Power Cores are the game's premium items and you can purchase more of these in the App Store. You also earn in-game money from doing jobs. When you encounter new ghosts, you research them. This takes time, so you can use Power Cores to speed things up. Research unlocks new items and power-ups for your team.

When it comes to fighting the ghosts, the game reverts to a side view, in which you move your team into position around the ghosts. One member is the healer of your team, another the wrangler, who's a defensive unit, and the last one is the one with the proton pack to capture the ghosts.

THE LOWDOWN

The music is excellent, as it's lifted straight from the first movie and the cartoony graphical style makes each character identifiable with their real-life versions. The sound effects are also taken directly from the movies, so you get the real Ecto-1 siren and the sound of the proton packs firing.

Ghostbusters makes good use of the licence: the map and busting scenes are well drawn, and we always enjoyed unlocking new levels and equipment. The game has all your busting needs covered, so you won't need to call anyone else.

You Might Also Like
If you like this style of game, you'll enjoy The Simpsons: Tapped Out (page 234–35).

Hot Tip
Don't cross the streams!

GREAT LITTLE WAR GAME

Category: Strategy • **Available on**: Android/iOS • **Cost**: Free ·
Download size: Small • **Age rating**: 9+

SYNOPSIS

Dominate the battlefield in this quirky little title that packs a big punch.

Great Little War Game is a turn-based strategy game where you and your opponent simply take
it in turns to wage war on one another. It sees you in command of a squad of troops over a
series of missions. The missions are varied in purpose, but all contain the same elements,
meaning you are essentially grinding the opposing force into a fine paste.

GAMEPLAY

You each get a turn at moving your men around the game world. Each man has a limited number
of moves they can make, and the game board is subtly divided into a series of circles that count
as spaces in which each character can move. To move them, you tap the desired character on the
head and then tap where you want them to go. If you decide that you're not happy with where

you moved them, you can reposition them
before your turn ends. If you continue to
press your man, the game will show you how
far they're able to move and how far they're
able to fire. Some units will have a short
range, some will have a longer range and
others are specialists that can't fire at all.

Once you've aligned your forces, you press
the play button that ends the day and begins

the opponent's turn. If your troops die, you can buy new ones from your base with money you earn each day. Ideally, you want to complete each level in as few days as possible to get the best score.

THE LOWDOWN

Aside from the chunky campaign mode, there's a pass 'n' play option where you and a friend share the device in games against each other. Plus there's Skirmish mode, where there are no mission parameters, just all-out carnage.

The game is viewed from the top down, but there's a lovely level of depth that really shows off the game's 3D capabilities. The animation on your little guys, as they run and manoeuvre through the different terrains, is terrific. When your boys are engaged by the enemy, or vice versa, the game zooms in on the troops as they trade fire. The action is fun, and the mission structure makes the game easy to play in whatever time you have.

You Might Also Like
Great Big War Game includes a huge campaign and new units to play with.

Hot Tip
Try to always have two snipers, and use them to flank the enemy on higher ground.

GUNS 'N' GLORY

Category: Strategy • **Available on:** Android/iOS • **Cost:** Free/Paid
Download size: Small • **Age rating:** 12+

SYNOPSIS

Tower-defence gets a new twist as you relieve settlers of their cash and worldly goods in the Old West.

In Guns 'n' Glory, you're in charge of a group of outlaws that sets up on top of canyons through which settlers are passing. Your goal is simply to stop them. So far, it sounds like a standard tower-defence game, with you placing your men in strategic positions around the canyons to take out the settlers. There are different classes of outlaws you can hire, from the standard Desperados through to Natives, Markswomen, Banditos and others. These characters are dotted around the map fast asleep, and you start off with just a couple to work with. As you start destroying settlers, you begin bringing in the cash. When you've accumulated enough cash, you can wake up the sleeping outlaws and pay them to join your gang.

GAMEPLAY

Where Guns 'n' Glory differs from other tower-defence games is that here you can reposition your men wherever and however often you want. As the game begins introducing multiple paths through each level, you have to manoeuvre your men around the map to cover the different routes.

Control of your outlaw gang couldn't be easier. To move one of your little guys, you just touch him and drag him in the direction you want him to go. You release your finger when he's reached your desired destination, and the game also allows you to choose which direction he begins aiming in.

THE LOWDOWN

Guns 'n' Glory also offers a good range of territories (six in all, each consisting of 10 stages) from around the United States, ranging from Mississippi in the Deep South all the way up to Alaska. That's a terrific range of levels to get through and, whilst the aim of the game never varies, each territory offers a welcome change of scenery.

The game has a Western soundtrack, which really suits the setting. The cartoon-style graphics fit well and the entire game has its tongue firmly in its cheek. Guns 'n' Glory is sufficiently different from other tower-defence games to be worth a look.

You Might Also Like
Guns 'n' Glory: Heroes
puts a fantasy spin on
the franchise.

Hot Tip
The natives are useful for setting
fire to wagons, so put them close
to the start line to soften up
the convoys.

JELLY DEFENSE

Category: Strategy • **Available on:** Android/iOS • **Cost:** Paid
Download size: Medium • **Age rating:** Suitable for all

SYNOPSIS

Defend your green crystals from the jelly invaders in this cute tower-defence game.

Jelly Defense doesn't really do anything new with tower-defence game mechanics. In fact, it simplifies things slightly by giving you preset positions on the map where you can place your towers. What Jelly Defense does do, however, is give you a beautifully presented and thoroughly enjoyable version of the genre.

The art style of this game is outstanding and we absolutely loved it. The developer has made lots of different jelly titles using this style and it works wonderfully. The map is effectively grey and white, but that description sells short the work that has gone into shading and drawing the environment.

GAMEPLAY

The invading jellies are two colours, red and blue, which corresponds with the colours of the towers used to attack them: blue towers attack blue jellies, red towers attack red jellies, and there's a two-tone tower that attacks both.

The jellies want the green crystals located at the end of the map and they follow set paths through the levels in their attempts to get them.

You set your towers up on the available spots along the path and let the jellies have it with all you've got. The destroyed jellies then drop orange globes, which you need to collect to buy more towers or to upgrade towers when things get a bit frantic. You can also sell a tower and buy another one if you fancy changing your strategy. Power-ups also appear from time to time and you need to use these tactically to get the best out of them.

THE LOWDOWN

The music in Jelly Defense is particularly impressive, and in fact it's so impressive that it's available to buy on iTunes. Later on, little splashes of colour are added to the levels such as on a waterfall or a tree. These small visual flourishes are fantastic and the bigger the screen you have, the better the game will look.

The touchscreen controls are perfect and the manipulation of the towers is very easy and intuitive. Jelly Defense's presentation will certainly appeal to kids, but don't underestimate this cute little game. It's a real challenge and will last you for a long time.

You Might Also Like
Jelly Band keeps the same visual style as Jelly Defense, but asks you to create your own musical ensemble instead.

Hot Tip
If you can, try to place a few towers behind the crystals, just in case some jellies break through.

JURASSIC PARK BUILDER

Category: Simulation • **Available on:** Android/iOS • **Cost:** Free (with in-app purchases) • **Download size:** Medium • **Age rating:** Suitable for all

SYNOPSIS

Build your own theme park full of live prehistoric exhibits. Just make sure you keep them locked up.

The idea of a theme park or zoo full of dinosaurs isn't such a bad one, although things went horribly wrong in every *Jurassic Park* film. Having said that, a park full of salad-eating dinosaurs might be a better idea than one that struggles to contain the sort that likes its food to run. With Jurassic Park Builder, you can do exactly that: you get to build your version of Jurassic

Park, however you like. You start off with a dense forest island that's pretty much a blank canvas for you to do whatever you like with. You are also given a visitor's centre, a lab, docks to bring in supplies and just the one dinosaur, a Triceratops.

GAMEPLAY

The game has an isometric view of your park and you navigate it by dragging the screen in the direction you want to go. Everything is easy to understand and the game leads you through the first couple of actions to allow you to get used to the style of play and menus.

Alongside the free play, there are also missions you can complete to earn more money, more dinosaurs and more visitor facilities. These missions are given to you by recognizable characters from the three movies. As you clear the land surrounding your park, you discover amber, which helps you research different dinosaur species, and soon you'll be well on your way to having 34 different species running around your park.

THE LOWDOWN

The use of the Jurassic Park licence is impressive and the detail in your park is all significant and intricate. The realism even means that if you stop feeding your exhibits, they'll soon start breaking out and finding other things to eat, so the management of your park is just as important as building high-security pens for your Velociraptors.

If you want to speed things along, you can buy stacks of cash from the App Store or Google Play. However, we thought we built a rather impressive park without doing so. The game is a fine addition to the genre, although maybe it's not a good thing that you want to see your dinosaurs escape, just to see what happens.

You Might Also Like
The Simpsons: Tapped Out (pages 234–35). If running a dinosaur theme park sounds too daunting, hang out with the residents of Springfield instead.

Hot Tip
Build plants and flowers around dinosaur exhibits to give you a coin boost for that dinosaur.

LITTLE EMPIRE

Category: Simulation • **Available on**: Android/iOS • **Cost**: Free (with in-app purchases) • **Download size**: Small • **Age rating**: 9+

SYNOPSIS

Take on the world as you try to expand your own little empire.

You can't beat a bit of world building. There's no greater satisfaction than taking a bare patch of land and turning it into a thriving community. Little Empire has a very strong social aspect that encourages you to make friends with other players, talk to them and, the most fun aspect, attack them and steal their stuff. In fact, the game uses your actual geographical location to determine how long it will take you to raid other people.

GAMEPLAY

You start off with a small kingdom with a castle and not much else. You're asked to pick a hero to lead your troops into battle. As the game progresses, you can upgrade him with armour and magic. However, the game doesn't abandon you to this world straightaway. Through the first few minutes

of play it will tell you all you need to know about how to set things up. These come in the form of tasks, which stay with you throughout the game and you'll keep referring to them. After a while though, you'll only refer to the tasks in the downtime between building things and waging war on other people.

Building something new or adding extra troops takes a bit of time and you can speed this up with Mojo. Mojo can be earned every time you level up, but it can also be purchased with real money if you want. There's no requirement to do so and, as usual with these types of games, not buying more doesn't harm your enjoyment.

When you go into battle, you enter the arena, a series of squares with you on one side and the opposition on the other. The goal is to beat up your opponent's army and reach their castle. Your hero is in the arena and can cast magic spells to give you an edge.

THE LOWDOWN

The game might be called Little Empire, but it is huge, and with its online and social-networking features, you'll never get bored. Soon you'll be checking in all the time to see if anyone has stolen your gear and if they have, you'll be waging a war to reclaim your stuff and take some prisoners. It's deeply involving, nicely presented and easy to get into.

Hot Tip
Try to get your crystal production into hourly cycles. It will earn you more.

MAJESTY: FANTASY KINGDOM

Category: Simulation • **Available on:** Android/iOS • **Cost:** Paid
Download size: Small • **Age rating:** 9+

SYNOPSIS

Rule your kingdom as you see fit. Just don't expect your subjects to listen to you.

Majesty: Fantasy Kingdom first appeared on PCs back in 2001. Now it's on mobile platforms, allowing you to take your kingdom with you wherever you go. The twist in Majesty is that, although you're the ruler and overseer of your kingdom, you can't actually tell people what to do.

GAMEPLAY

Usually with games of this type, you have your different characters, whom you'd send off on different quests to defend your kingdom and raid goodies. Here, it works differently. You can't order your knights, wizards or anyone else to do anything. Instead, you have to incentivize them, be it going on quests to kick out unruly goblins, or scouting unexplored parts of the map.

You do this by offering a reward for the tasks and seeing if it takes the heroes' fancy. The trade-off for this is that you need to create a thriving economy to make sure that your heroes spend their money in your markets or blacksmiths. So you get the money back, just in a roundabout way.

The detail on the individual buildings you construct in your kingdoms is very good and we loved the oversized animation on the human characters as they wandered about. The heroes are individuals and have their own names and levelling-up system. The upside is that all of these warriors can become impressive allies; the downside is that when one of them dies, they're gone for good and you have to make do with weaker heroes to replace them. You can get quite attached to them. As your kingdom gets bigger, you'll start to

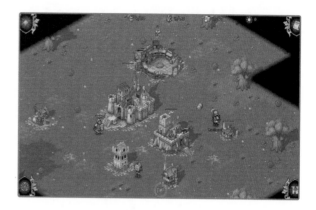

incorporate different races and warrior classes like elves and barbarians, which is welcome when enemies begin encroaching on your kingdom.

THE LOWDOWN

The navigation buttons to bring up the menus are nicely mounted on each corner of the screen, meaning you get a good view of the map. On most screens, the menus are very easy to navigate, but on older, smaller screens the text boxes are a little small, so be aware of that. Obviously, on tablets that isn't an issue and the whole thing looks fantastic.

We enjoyed the old-school feel of Majesty and especially liked the fusion of real-time strategy and resource management. They just don't make them like this any more.

You Might Also Like
The developer, HeroCraft, also brought us Ant Raid (pages 194–95), a wacky take on real-time strategy.

Hot Tip
You can use magic to reveal unexplored map areas, but rangers can also do this for you.

MINECRAFT – POCKET EDITION

Category: Simulation • **Available on:** Android/iOS • **Cost:** Free/Paid
Download size: Small • **Age rating:** Suitable for all

SYNOPSIS

The home computer smash hit is now available on mobile devices. Be prepared to expand your world.

First thing, don't be put off by how the game looks. It's intentionally designed this way, as it makes crafting in the landscape easier. Plus part of the charm is its old-fashioned, early 1990s PC style. Anyone who has ever used a pre-millennium version of Windows will instantly recognize it.

GAMEPLAY

Minecraft drops you into a randomly generated environment and lets you build anything, and we do mean anything. The only limit here is your imagination: you have the tools, but do you have the talent? We admit that to some people the prospect of being dropped into an empty virtual world and being allowed to do whatever you want can be daunting. It's certainly unusual for games to do that these days, but Minecraft trusts you to work things out for yourself and use your imagination.

The game is controlled via an on-screen pad on the left of the screen. You start with a first-person view, but that can be changed to a third-person perspective if you prefer (we did). Creative mode allows you to build whatever you want, however you want. That's it. You build, you explore and you build. This is the game's more artistic side. Survival mode is the gaming side. Here, you have to look for food, build shelter and look out for monsters that come out at night.

You break, or mine, the area to pick up items and resources that you can use to build elsewhere. You place a block from your inventory (there are lots of different materials) wherever you want and from there you can craft whatever you like. Hence the name.

THE LOWDOWN

If you're not sure about Minecraft, then we recommend the free version. It'll get you into the game and is only limited by its lack of a save function. If you find that you're

converted, then the paid version has more building materials and a save function, so you can get to work building your own house, or maybe a park.

Minecraft is gaming in its freest form and once you start building, you'll be hooked. It may look basic, but this is one of gaming's most sophisticated titles.

You Might Also Like
There really is nothing else like Minecraft. Maybe Lego Star Wars: Battle Orders, but only because they both have blocks.

Hot Tip
Gravity isn't in play with Minecraft, so use that to your advantage.

PARADISE ISLAND

Category: Simulation • **Available on:** Android/iOS • **Requires:** iOS 4.3+/
Android 2.1+ • **Cost:** Free (with in-app purchases) • **Download size:** Small
Age rating: Suitable for all

SYNOPSIS

Build your perfect tropical island getaway with a game that guarantees you won't get sunburnt.

Paradise Island tasks you with building the ultimate holiday destination. You start off with a bare beachfront resort, but soon you'll have built so much you won't be able to see the sand any more. That's a good thing, right?

GAMEPLAY

At the start, all you have is your main hotel building and a broken-down pier. Your first task is to fix that up to get the tourists in. From there, it's all about making people happy. This means you can build anything from gift shops to water slides, with some wacky designs in between.

Although the idea of the game reminded us of Theme Park, and its graphical style definitely had nods to the Tycoon series, Paradise Island is more similar to The Simpsons: Tapped Out (see pages 234–35). Adding new buildings and upgrading them gets you experience points (XP points) and cash. You need these to get you to the next level; every level you open grants you access to new items to purchase for your island. As with The Simpsons, there's a premium currency in the game. In Paradise Island's case, these are called Piastres. They speed up tasks and buy you premium items; if you want more, they're available to buy with real money. As with nearly all premium items in these situations, they're not essential and the game is just as enjoyable without using them.

THE LOWDOWN

The interface is easy to use, with all of your inventories and tasks laid out at the bottom of the screen. To collect money or XP from a building, just tap the building and the items go straight into your bank. Adding a new building is similarly easy, as you select it from the inventory and place it where you wish. As befitting a game called Paradise Island, the game has a bright and colourful graphical style. The music is minimal and you can even hear the soft crashing of waves against the shore underneath.

It's surprising that this idea has never been done before, but rest assured that Paradise Island does it very well. You'll enjoy the ability to dip in and out of it, and soon you'll have your island paradise looking like the Dubai of the digital world.

You Might Also Like
Try a tropical island with a difference in Jurassic Park Builder (pages 216–17).

Hot Tip
Early on, plant five flowers and double your cash and XP on them.

PLAGUE INC.

Category: Strategy • **Available on:** Android/iOS • **Cost:** Free/Paid
Download size: Small • **Age rating:** 9+

SYNOPSIS

Star in your very own disaster movie, except this time, instead of saving humanity, you're trying to make it extinct.

Plague Inc. asks you to create a disease and make it as infectious and lethal as possible. If you're successful, then humanity will be wiped out. Whilst we agree this doesn't sound like the most inviting premise for a game, you'll have to trust us when we say that underneath this is a clever, innovative management game that subverts your instincts about disease and survival.

GAMEPLAY

You start off by picking what form your virus will take. You start with bacteria, but you access different pathogens as you eradicate society. The game screen is a map of the world. Select a country to either infect with your disease or to see how the population is coping with extinction in the face of a current contagion.

As you go through the game, bubbles pop up on screen. Red bubbles mean a new infection in a new country, yellow bubbles mean you've earned DNA points and blue bubbles mean a cure is being developed. As they pop up, you have to tap them to get DNA points to evolve your disease and, in the case of the blue bubbles, stop the cure. The DNA points that evolve your disease are used to decide by which methods it will be transmitted, such as by birds, rodents, cattle, blood, air or water.

DNA points also dictate what abilities your disease has: whether it can survive in the cold or heat; whether it can resist drug treatment; and whether it has a bacterial shell to protect it in all environments. The game will ask you to choose the symptoms, a rather gruesome task. You can mutate it here, too, so what might start out as a cough could lead to heart problems, which could be later linked to immunity suppression.

THE LOWDOWN

If you have the paid game, you can speed things up and modify your disease's genetic code at the start to further customize your plague. It's a nice addition if you're feeling particularly infectious.

What, on the face it, seems like an odd subject matter is actually a very clever tactical game that becomes a global version of cat and mouse between you and humanity. At the very least, you'll start washing your hands a lot more.

You Might Also Like
If apocalypse is your style, try Plants vs. Zombies (pages 228–29).

Hot Tip
Transmission of your disease is the key to winning. Get the disease to everyone before making it lethal.

PLANTS VS. ZOMBIES

Category: Strategy • **Available on:** Android/iOS • **Cost:** Paid
Download size: Small • **Age rating:** Suitable for all

SYNOPSIS

There's a zombie apocalypse going on and mankind's best defence is ... plants?

Plants vs. Zombies has to be one of the best-known and popular games for mobile platforms. Its name alone stands out, but once you start playing, it's the mix of tactics and strategy that will keep you enthralled.

GAMEPLAY

The setup is that zombies are attacking your house. Luckily, you have something up your sleeve that the zombies haven't planned for: plants. Lots and lots of plants, 49 different types of zombie-smacking plants, to be precise. You start off with a blank front lawn outside your house. Across the street, there's a zombie convention and they've decided they want your tasty brains. Obviously you aren't going to stand for that, so you have to deploy different types of plants on your front lawn to keep the zombies at bay. You do this by selecting a plant from the list on the left of the screen and placing it anywhere you want on the lawn.

You start off with a fairly standard roster of plants, but as you progress through adventure mode, you get more and more plants, of various types, each type capable of different things. With so many varieties, we can't list them all, but among our favourites were Snow Peas, which fire frozen peas at the zombies, and Grave Busters, which destroy the graves that have popped up on your lawn.

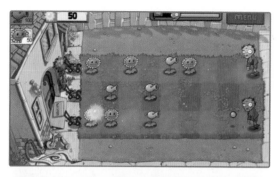

No plant is essential bar the sunflower, which generates sun rays. The more sun rays you have, the more plants you can buy to stave off zombie domination. As you progress through the game, you can select which plants you want in your inventory, and with 49 to eventually pick from, you'll always have a wide selection.

THE LOWDOWN

The graphics are bright and cheerful, and the animation on both plants and zombies is very well done. The music is extremely catchy too, with a nice balance between bouncy and macabre. If you don't want to play the adventure mode, there's also a quick-play option. The game does save your adventure mode though, and you can add separate adventure campaigns if there's a couple of you sharing a device.

There's plenty on offer in Plants vs. Zombies, including mini games, upgrades and hidden codes. It all adds up to a substantial package worth anyone's money. You should have this game.

You Might Also Like
Peggle, another game from Popcap, is completely different, but just as much fun.

Hot Tip
Always start with at least four sunflowers to give you a boost.

REBUILD

Category: Simulation • **Available on:** Android/iOS • **Cost:** Paid
Download size: Small • **Age rating:** 12+

SYNOPSIS

Society has collapsed and the world is overrun with zombies. Gather survivors, fight back against the zombie hordes and rebuild your city.

Rebuild takes a premise that has been done to death (pardon the pun) and takes a new approach that's both surprising and enjoyable. If you can call living out a zombie apocalypse enjoyable, that is.

GAMEPLAY

You start by naming your character and town, deciding what season it is (winter being harsher than spring), what weapon your character has and how much of a struggle taking the city back was (the difficulty). Once you've got all that done, it's time to survey your post-apocalyptic, zombie-infested city. Early priorities include setting up guards, growing food on farms and scavenging for items in buildings close to your fort. The game even goes as far as to add human gangs that will attempt to steal your supplies and generally cause trouble.

You can assign different characters different tasks as you go about reclaiming the areas around your fort. One of the cleverest aspects of Rebuild is that the map is randomly generated every time you start a new game, so no two games are alike. Interacting with Rebuild is easy, as everything you need to know is displayed in your status bar at the top of the screen. This bar includes how much food you have, how susceptible you are to zombie shenanigans and how happy (relatively speaking) your people are. There's also an inventory menu on the right of the screen that gives you an overview of your city.

Rebuild runs in days, tasks take a few days to complete and when you've completed your day's tasks, you just press the button at the top of the screen to end the day. The next day starts immediately and includes updates if anything has occurred overnight, such as survivors turning up or people asking to trade with you.

THE LOWDOWN

It's surprising that this type of scenario hasn't been applied to this genre before, but it's well executed and there's a great soundtrack that really captures the mood. The amount of gaming you'll get out of this, coupled with its random maps, means that the price is an absolute steal and we had immense fun playing it.

You Might Also Like
Zombies (again!) in Plants vs. Zombies (pages 228–29), one of the funniest games out there.

Hot Tip
To begin with, you might want to keep your band of survivors small, so there's less demand for food.

RESTAURANT STORY

Category: Simulation • **Available on:** Android/iOS • **Cost:** Free (with in-app purchases) • **Download size:** Small • **Age rating:** Suitable for all

SYNOPSIS

You have the chance to run your very own restaurant in this cute management simulator.

If there's one thing everyone loves, it's food. You can't beat good food, and good food in great surroundings is even better. In Restaurant Story, you have the opportunity to design those surroundings in your own restaurant. The idea with Restaurant Story is you have to build up your restaurant from meagre beginnings into the kind of place you'd have to wear a tie to get into.

GAMEPLAY

You start off with a very small surface area, only big enough to fit two cookers and three tables. Immediately, customers start coming in and you must serve them food and keep them happy. Initially, the food you're offering is rather meagre, salads and omelettes, but the happier your customers are when they leave, the more money and recipes will be available to you. You do this by selecting them from your cookbook that pops up above the stove. To alter the décor, you go into your design menu where there are hundreds of different items to buy with the money you've earned.

The food in Restaurant Story cooks in real time. So for instance, it takes five minutes to prepare an omelette. Later on, the more difficult dishes take a lot longer, so you can speed things up by spending gems if you wish, which are available to buy in the respective app stores. If you want to wait though, there's plenty to do and the game can be left to its own devices whilst you're busy elsewhere.

There's a great social aspect to Restaurant Story, as you can visit other people's restaurants and rate their efforts and, of course, they can do the same to yours.

THE LOWDOWN

We liked the clear presentation of Restaurant Story: the first thing that strikes you about this game is its graphics; the characters are large; all the items that you cook or buy are clear and the whole package is very easy on the eye, plus it has a bouncy soundtrack.

You can start your restaurant off and then drop in and out of it whenever you need to, so you're able to dedicate as much time to it as you wish. It's definitely more filet mignon than cheeseburger.

You Might Also Like
Zoo Story 2 (pages 240–41) is from the same company, but instead of serving up animals, you're looking after them.

Hot Tip
To attract more wealthy customers, make the décor in your restaurant as pleasing on the eye as possible.

THE SIMPSONS: TAPPED OUT

Category: Simulation • **Available on:** Android/iOS • **Cost:** Free (with in-app purchases) • **Download size:** Medium • **Age rating:** Suitable for all

SYNOPSIS

Homer's caused a meltdown (again!) and it's up to you to rebuild Springfield with the aid of all of your favourite Simpsons characters.

Sometimes, the best fun you can have is with family, and with the Simpsons we all know them so well that they practically are. The Simpsons: Tapped Out is very simple to play, but incredibly hard to put down. It sees you in charge of rebuilding Springfield after Homer causes an accident by spending too much time playing with his MyPad. You start off by building the Simpsons' home with Homer and Lisa, and from there you can begin repopulating your version of Springfield.

GAMEPLAY

This is done by earning money and experience points (XP) when you add new buildings, buy more land or even send characters to the Kwik-E-Mart. The more XP you gain, the more quickly you move to the next level. With each new level come more characters, more buildings and even plants, trees and amenities for your town. Each character has a specific quest, which means you'll always have something going on. Whether it means sending Homer to lounge in his pool or Moe to smuggle endangered species, there's always something to occupy your interest.

If you want to make things go a bit faster, or you just want to buy some exclusive items, then you can do so by using the in-game currency of donuts. You earn these by moving up through levels and by completing specific quests. If you feel like you want to move things along even more quickly, you can buy donuts or even purchase an Itchy and Scratchy scratch card to win more. You can also store items that you've bought for later use in your inventory, which comes in handy when you decide that you don't want a Krusty Burger next to the beautiful park you've built.

THE LOWDOWN

The game couldn't be simpler to pick up and play, the emphasis throughout being the

ability to play using one finger (hence the name). All your favourite characters are here and presented in beautiful HD-quality graphics, which make you feel like you're controlling your very own episode of *The Simpsons*.

You Might Also Like
**Jurassic Park Builder (pages 216–17).
If the thought of hanging out with
the Simpsons is too safe, try playing
fetch with a Tyrannosaurus.**

Hot Tip
**Tap Homer 10 times in a row to
get a free Jebediah Springfield
statue and 10 free donuts.**

THE SIMS: FREEPLAY

Category: Simulation • **Available on**: Android/iOS • **Cost**: Free (with in-app purchases) • **Download size**: Medium • **Age rating**: 12+

SYNOPSIS

Create your own living, breathing world with EA's smash-hit simulator of life.

The Sims has been a real success story for EA. Not just because it's immensely popular, but also because its appeal is so broad. The Sims probably appeals to the widest demographic in all of gaming. Everyone can play the Sims and it's brought gaming to the masses like nothing else.

Not satisfied with conquering the home, The Sims is now available on mobile devices. EA have been even more generous this time around with The Sims: Freeplay, as now you can enjoy the Sims experience without paying a thing. The same principle that's used in games such as The Simpsons: Tapped Out (*see pages 234–35*) is available here. Setting your Sims tasks or building new houses plays out in real time, but if you don't want to wait, you can buy life points to speed things up. The choice is entirely yours.

GAMEPLAY

If you're not familiar with The Sims, then all you need to know is that you control the Sims of the title. Sims are virtual people and you can do almost anything with them. In terms of how it works on touchscreen, The Sims actually feels better than it does on home computers. The whole interface is easy to navigate and, whilst it does feel overwhelming at the start, the game leads you by the hand until you're comfortable. Soon you'll be setting your Sims tasks, making friends and running your new social environment.

The game sets you missions to accomplish. Do that and you'll have money to spend on ... well, anything.

The number of customization options on offer is absolutely staggering, from what type of carpet you want to what colour socks you want to wear and much, much more. You'll never tire of tweaking the little things in The Sims.

THE LOWDOWN

The Sims is a big game, both in terms of options and permutations, but also graphically. The Sims' world is fastidiously created in 3D and it looks fantastic. The Sims is worthy of your time, and once you're in, you'll never want to leave.

You Might Also Like
The Simpsons: Tapped Out (pages 234–35) is a slightly different take on the Sims formula.

Hot Tip
When you get your dog, have him dig stuff up. When he digs up life points, make sure you praise him. He'll then start developing a taste for digging up more.

STEAMBIRDS

Category: Strategy • **Available on:** Android/iOS • **Cost:** Paid
Download size: Small • **Age rating:** Suitable for all

SYNOPSIS

Take to the skies to destroy the enemy in this exciting turn-based combat game.

If you've ever fancied yourself as an ace fighter pilot or even as a general in the field, then we've got the game for you. SteamBirds is a turn-based game where you control squadrons of planes in an alternative aviation history where steam is king. If you're not sure what a turn-based game is, it's exactly as the name suggests. In most games, notably home computer formats, it's a case of you taking your turn, then your opponent taking theirs, kind of like the video-game equivalent of a board game.

GAMEPLAY

SteamBirds isn't quite like that. Here you plan your moves and press the forward button. The action then moves along for a few seconds before pausing for you to plan your next move.

That's done by dragging the on-screen arrows in the direction you want your aircraft to go. You don't have to worry about firing, the game does that for you, and that's not really the emphasis of turn-based combat games anyway.

There are other moves you can make, like speed your planes up or have them perform quick turns. You soon realize that you're taking part in an airborne ballet and it's actually very graceful as you try to outmanoeuvre the enemy and send them plummeting to the ground.

You view the maps from the top down and the game has a very retro feel that suits it well. The maps look like those you'd imagine spread out on the table in Churchill's wartime bunker and the effect really enhances your enjoyment as you play. The game has a set of main levels and two lots of bonus levels. As you progress, you're given star ratings based on how much damage you sustained during your mission, and you're briefed before you start each mission.

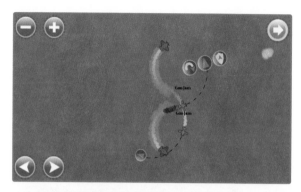

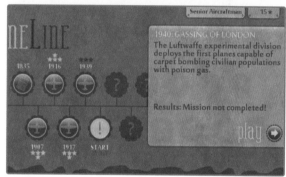

THE LOWDOWN

It looks basic, but SteamBirds deserves a chance to impress you. The music is very rousing and evocative of the era of aviation it's trying to convey. Once you down your first enemy and see them spinning to the ground, you'll be hooked. Chocks away!

You Might Also Like
Try a different sort of flying game with Flying Penguin. It's a complete change of pace, but it's fun, and penguins can fly, right?

Hot Tip
Use the boost option to catch up with fleeing enemies, but never use it for a head-on assault.

ZOO STORY 2

Category: Simulation • **Available on:** Android/iOS • **Cost:** Free (with in-app purchases) • **Download size:** Small • **Age rating:** Suitable for all

SYNOPSIS

It's nature versus nurture as you get to control your own zoo with this business sim that's smarter than the average bear.

Borrowing the template from other TeamLava games, Zoo Story 2 asks you to manage your own zoo. It's very simple to use and the interface has been designed so anyone can just pick it up and play. Of course, building and maintaining your own zoo is a tricky business. If it's not the animals, then it's the people who need looking after, so you have to provide food and drink stands, as well as benches, ponds and other things to make your zoo look pretty.

Making your zoo look attractive isn't difficult when the graphics look as good as they do here. There's no realism on offer, as the animals are all rendered in a delightful cartoon fashion with chunky features, and each and every one of them looks extremely cute, from the lions to the frogs.

GAMEPLAY

You earn coins to improve your zoo by adding animals and amenities. Everything is one touch and easy to navigate. You're also set tasks to get your zoo up and running, one of which is to visit other people's zoos. This is a terrific way to get used to the online community that you are encouraged to be involved in.

The most fun and interesting aspect of Zoo Story 2 is breeding your animals. Tap an animal to buy that animal a mate. They then breed for more animals; these will earn more coins, as your customers love seeing a family of animals.

THE LOWDOWN

So far pretty normal, but here's where Zoo Story 2 is quite innovative. You can crossbreed animals to make new species or bring back animals from extinction. For example, crossing a polar bear with an elephant will get you a mammoth. It's great fun for experimenting and brings a new dimension to the game.

This extra innovation alone would make Zoo Story 2 worthy of your time, but we loved the cartoony visuals and buoyant calypso music that accompanies gameplay. If it's realism you're after, then you've come to the wrong place, but if you want a fun game that will pass the time and make you smile, then Zoo Story 2 will leave you extremely satisfied.

You Might Also Like
Anything in the Story family by TeamLava is fun, but try Jurassic Park Builder (pages 216–17) for a management sim with more bite.

Hot Tip
Fancy a mythic animal for your zoo? Crossbreed a big-horned sheep and a gorilla to get a minotaur.

ALTERNATIVE TOP 20 LISTS

ALT TOP 20 RETRO GAMES

Just for fun, here are Flame Tree's views on some alternative top 20s. You'll recognise many from the book and we've thrown in a few extra popular games too, that we just couldn't miss off of these lists. The following list of retro games will have you revelling in the nostalgia of yesteryear,

Above: Everyone loves the classic game Tetris.

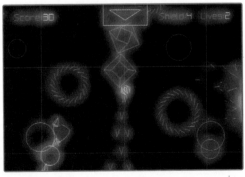

Above: PewPew is inspired by Atari's Asteroids.

1. FROGGER

2. PAC-MAN

3. SONIC CD

4. TETRIS

5. DUKE NUKEM 3D

6. SUPER BIT DASH

7. TINY TOWER

8. PEWPEW

9. PINBALL ARCADE

10 DRAGON'S LAIR

11 DOUBLE DRAGON

12 METAL SLUG 3

13 SNAKE '97

14 SIMPLE BRICK BREAKER

15 SPACE INVADERS INFINITY GENE

16 MEGANOID

17 RADIANT

18 NBA JAM

19 STARDASH

20 SYDER ARCADE HD

Above: Syder Arcade HD is a great space-based shoot 'em up game. A throwback to games of the 1990s, it is perfect for any fan of old-school shooters.

Above: NBA Jam was a console classic back in the 1990s, and still just as enjoyable.

ALT TOP 20 MOVIE & TV GAMES

From fast-paced superhero action to cartoon capers, your favourite movies and TV programmes take on a life of their own in Flame Tree's alternative top 20 list.

Above: Rebuild Springfield after Homer causes a meltdown in this game featuring all of your favourite Simpsons characters.

Above: You can build your own version of Jurassic Park.

1 THE HOBBIT: KINGDOMS

2 THE SIMPSONS: TAPPED OUT

3 MARVEL WAR OF HEROES

4 FAST & FURIOUS 6: THE GAME

5 TRANSFORMERS LEGENDS

6 JURASSIC PARK BUILDER

7 WRECK IT RALPH

8 IRON MAN 3 – THE OFFICIAL GAME

9 SMURFS' VILLAGE

Above: Who you gonna call?

Above: Kids and grown-ups will love Toy Story: Smash It!

Above: Put your movie knowledge to the test!

ALT TOP 20 ANIMAL GAMES

From penguins that think they can fly to running your own zoo, Flame Tree has chosen 20 animal-based games that are bright, brilliant and pure unadulterated fun.

Above: Angry Birds is the most popular touchscreen game ever and with good cause: who knew smashing pigs with birds was such fun?

Above: Maintaining your own zoo is a tricky business.

1 ANGRY BIRDS

2 WHERE'S MY WATER?

3 SONIC THE HEDGEHOG

4 WORMS 2: ARMAGEDDON

5 FLYING PENGUIN

6 ZOO STORY 2

7 GO GO GOAT!

8 BIRZZLE

9 WHALE TRAIL

Above: You can't get more fun than monkeys in balls.

Above: Angry Birds from the opposite point of view.

Above: Eager Beaver is a challenging strategy game.

ALT TOP 20 ADDICTIVE GAMES

There are certain games that you just can't put down or they will drive you crazy until you complete them. Here are Flame Tree's alternative 20 addictive games.

Above: An Indiana Jones-inspired adventure.

Above: Fed up of your five a day? Get your revenge.

1. CANDY CRUSH SAGA

2. DOODLE JUMP

3. ANGRY BIRDS

4. CUT THE ROPE

5. TEMPLE RUN

6. FRUIT NINJA

7. PAPER TOSS

8. ROBOT UNICORN ATTACK

9. WORDS WITH FRIENDS

Above: Beat the zombie apocalypse with plants.

Above: Show off your artistic talents (or lack thereof!).

Above: Video games and art combine in a witty puzzle

FURTHER READING

Alger, K. and Smith, C., *iPhone Made Easy*, Flame Tree Publishing, 2013.

Baig, E.C. and LeVitus, B., *iPad for Dummies*, John Wiley & Sons, 2010.

Buckley, P., *The Rough Guide to the Best iPhone and iPad Apps: The 500 apps that your iOS device was born to run*, Rough Guides, 2012.

Clare, A., *The Rough Guide to the Best Android Apps: The 400 Best of Smartphones and Tablets*, Rough Guides, 2012.

Dickens, H.J. and Churches, A., *Apps for Learning: 40 Best iPad/iPod Touch/iPhone Apps for High School Classrooms*, Corwin, 2012

Fleishman, G., *Five-Star Apps: The Best iPhone and iPad Apps for Work and Play*, Peachpit Press, 2010.

Gookin, D., *Nexus 7 For Dummies*, John Wiley & Sons, 2012.

Laing, R., *iPad Made Easy*, Flame Tree Publishing, 2013.

LeVitus, B. and Chaffin, B., *Incredible iPad Apps for Dummies*, John Wiley & Sons, 2010.

McManus, S. and Hattersley, R., *iPad for the Older and Wiser, Get Up and Running with your iPad or iPad mini*, John Wiley & Sons, 2013.

Meyers, P., *Best iPad Apps: The Guide For Discriminating Downloaders*, O'Reilly Media, 2010.

Morris, K., *The Best iPhone Apps Ever: The Ultimate Guides to All the Apps Every iPhone User Needs*, CreateSpace Independent Publishing Platform, 2013.

Muir, N.C., *Kindle Fire HD For Dummies*, John Wiley & Sons, 2012.

Provan, D., *iPad 2 In Easy Steps*, In Easy Steps, 2011.

Rosenzweig, G., *My iPad (covers iOS 6 on iPad 2, iPad 3rd)*, QUE, 2012.

Schense, D., *What is a Tablet Computer? The Best Top 20 Tablets in Review: What you Need to Know Before you Buy*, CreateSpace Independent Publishing Platform, 2012.

Vandome, N., *Android Tablets in Easy Steps*, In Easy Steps, 2013.

Watson, L, *Teach Yourself VISUALLY iPad 2*, John Wiley & Sons, 2011.

WEBSITES

www.148apps.com

Provides insight into the best iOS applications through numerous reviews and latest news sections.

www.androidpit.com/en/android-market/recommendations

Thousands of app recommendations for android devices divided into categories such as newcomer, top downloads and price reduced.

www.appblogit.com

An exciting new blog that reviews apps and provides hints and tips to the beginner through tutorials.

www.appmodo.com

A game review and mobile applications blog covering applications, games, utilities, hardware and accessories.

www.guardian.co.uk/technology/appsblog

Frequently updated blog on all things app related from *The Guardian*.

play.google.com/store

Here you can buy Google apps, including thousands of games apps via Google Play for your Android device.

www.pocketgamer.co.uk

A site dedicated to reviews, news andsuggestions for games available on a wide range of different mobile devices.

www.store.apple.com

Thousands of apps available direct through the Apple Store.

www.theiphoneappblog.com

A blog committed to providing readers with unbiased and knowledgeable reviews of iPhone, iPad and iPod applications. Also offers written reviews, video reviews and a live audio show/podcast.

www.toucharcade.com

The largest site dedicated to iOS gaming, it covers the latest games and apps for Apple's iPhone, iPod touch and iPad.

INDEX